HEAD
SCRATCHERS
THE
PUZZLE
BOOK

Rob Eastaway is the advisor for *New Scientist*'s puzzle column. He is the author of several bestselling books on everyday maths, including *Maths On The Back of an Envelope, Why do Buses Come in Threes?* and *Maths for Mums and Dads*. He lives in London.

Brian Hobbs is the creator and host of the *Brain Drop Puzzles* podcast and a frequent contributor to the *New Scientist*'s puzzle column. He lives near Dallas, Texas with his wife and five children, for whom he proudly serves as maths homework consultant.

NewScientist

HEAD SCRATCHERS THE PUZZLE BOOK

ROB EASTAWAY • BRIAN HOBBS

ALLEN&UNWIN

First published in Great Britain in 2023 by Allen & Unwin, an imprint of Atlantic Books Ltd.

Copyright © Rob Eastaway, Brian Hobbs and *New Scientist* 2023

10 9 8 7 6 5 4 3 2 1

A CIP catalogue record for this book is available from the British Library.

Trade paperback ISBN: 978 1 83895 877 0
E-book ISBN: 978 1 83895 878 7

Printed and bound by CPI (UK) Ltd, Croydon CR0 4YY

Allen & Unwin
An imprint of Atlantic Books Ltd
Ormond House
26-27 Boswell Street
London
WC1N 3JZ

www.atlantic-books.co.uk

In memory of David Singmaster,
the great metagrobologist.

CONTENTS

3 All Things Considered
Puzzles featuring everyday objects

4 Figuring It Out
Problems that need some calculation

5 A Matter of Time...
Clocks, time and dates

6 Mind Games
Games with intriguing strategies

7 Eccentric Tales
Puzzles with peculiar characters

PART II
Solutions, Back Stories and Commentary ... 89

PART III
Hints 207

INTRODUCTION

Welcome to *New Scientist*'s first collection of puzzles for over 40 years. We've picked out 70 of our favourites from the magazine's popular weekly column, and grouped them into themes. You'll find puzzles that are great for sharing (or arguing over) with friends, problems drawn from real-life situations, games with intriguing strategies, and puzzles with such creative and amusing storylines that we couldn't help but share them with you. Regardless of the category, the puzzles are designed to entertain, enlighten, and intrigue, and we hope that you enjoy them as much as we do.

A particular feature of this book is that you'll get to peek behind the curtain and discover the previously untold backstories or concepts behind the puzzles, which are often interesting and entertaining in their own right. You'll learn why a particular puzzle adaptation involved talking to an expert in sheep genetics, the solution that was thought up by the BBC Radio 5 Drive team, and a variety of outside-the-box solutions to seemingly simple challenges. For this reason, the 'Solutions' section is more substantial than in other puzzle books, and is designed to be read more thoroughly rather than simply glanced at. Indeed we expect - and maybe even hope - that some readers will find this section at least as interesting as the puzzles themselves.

It wasn't easy to pick out only 70 of the puzzles from our collection. The list of wonderful *New Scientist* puzzles and the stories behind them is growing week by week. But, much like the 'Book of Numbers' that you'll discover on page 8, this book has to end somewhere. The good news is that we'll have plenty of material should we decide to produce a Volume 2.

PART 1

THE
PUZZLES

Chapter 1

PUB PUZZLES

Some puzzles lend themselves to being tackled with friends. We'll call them pub puzzles, but they work equally well in cafes, at a family lunch, on a train or in a canteen – anywhere that you can have a conversation around the table. You should be warned, however, that some of these puzzles can lead to extremely heated debate – even when you've read the solution.

1

CREATIVE ADDITION

ROB EASTAWAY

There is an old adage that one person's 'creativity' is another person's 'cheating'. This puzzle will test which side of the fence you sit on.

The numbers 1 to 9 have been written on cards and left on a table:

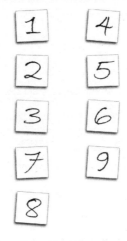

The left-hand column adds to 21, and the right adds to 24. Your challenge is to move just one card so that the two columns add to the same total. There's a classic 'aha' solution to this puzzle, but my daughter came up with a solution I wasn't expecting. Since then I've been offered at least ten more distinct solutions.

How many solutions can you find that you regard as creative rather than cheating?

2

THE H COINS PROBLEM

DAVID BEDFORD

Seven coins have been placed in the 'H' shape below. Altogether there are five lines of three, including the diagonals.

Your challenge is to place two more coins so that you can make *ten* straight lines of three. No stacking of coins, and no lines of four coins or other sneaky trick is required.

If you find a way to do this, give yourself a silver medal. If you find a second way to do it that isn't a mirror image of the first, award yourself a gold.

3

THE BOOK OF NUMBERS

HUGH HUNT

Polly plans to write a book (in English) containing all the whole numbers from zero to infinity in alphabetical order. She knows this will take her a very long time, but she makes a start. She figures that first on her alphabetical list is the number eight. After a while she tires of the task, jumps to the last page and starts working backwards. She reckons that the last entry will be zero.

Curiously, even though this book will take forever to finish writing, it's possible to state which number will be listed second in the book, and which one will be second-last. What are those two numbers?

(Note – when Polly wants to write numbers bigger than the quadrillions, i.e. numbers with fifteen zeroes, she strings numbers together, for example 'one billion trillion' or 'five million million quadrillion'.)

4

LATE FOR THE GATE

ROB EASTAWAY

This deceptively tricky everyday problem was first posed
by Fields Medallist Terence Tao in 2008.

You are in a bit of a rush to catch your plane, which is leaving from a remote gate in the terminal. Some stretches of the terminal have moving walkways (travelators); other portions are carpeted.

You always walk at the same speed, but your speed is boosted when you are on the travelator.

You look down and spot that your shoelaces have come undone. This won't slow you down, but it's annoying, so you decide to stop to tie them. It will take the same amount of time to tie your laces if you're on the carpet or on the travelator, but if you want to minimize the time it takes you to reach the gate, where should you tie your laces?

What if you are feeling energetic and can double your walking speed for five seconds? Is it more efficient to run while on a travelator, or on the carpet?

5

BUS CHANGE

KATIE STECKLES

I'm about to get on the bus, but the driver doesn't give change. The fare is £1 and I don't have exactly £1 in change on me, so I hand over more than £1 and they keep the change.

Once I've sat down, I realize that the amount of money I had with me was the largest possible amount I could have had in change without being able to pay £1 (or any multiple of £1) exactly. How much did I have?

(In case you need a reminder of British coinage, the coins are 1p, 2p, 5p, 10p, 20p, 50p, £1 and £2.)

6

DARTS CHALLENGE

ZOE MENSCH

On a regular dartboard the maximum that you can score with three darts is 180, by getting three treble 20s. However, there are scores below 180 that you can't get with three darts.

What's the lowest score you can't get with three darts? And for that matter, what is the lowest score that you can't get with two darts? And with one dart?

As a reminder, hitting the narrow outer ring doubles the score for that dart, and hitting the inner narrow ring triples it. Hitting the bullseye in the centre of the board scores 50 points, and hitting the ring that surrounds it scores 25.

7

EVENING OUT

PAULO FERRO

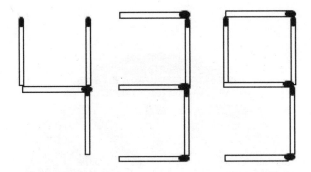

The figure above is composed of fifteen matchsticks. Move (but don't remove!) two matchsticks to different positions to get a 3-digit number in which all three digits are even numbers. There are three solutions (none of them 'tricks'). Can you find them all?

8

CAESAR CIPHER

ANGUS WALKER

How might Caesar get you from 3 to 47? A bit of general knowledge might help you here. Or a bit of numerology. Because surprisingly, there are two neat solutions to this puzzle.

9

SYMMETRIC-L

DONALD BELL

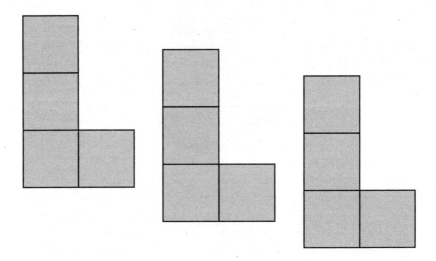

At first glance you wouldn't think it's possible to put these three L-shaped tetrominoes together to make a flat, symmetrical shape.

And yet... it turns out that there are *two* different ways to make a shape with mirror symmetry using *all three* Ls. If you can find them, you deserve an L-ympic medal.

To be clear, each L must be touching at least one other L and you are allowed to flip the Ls over, but no overlapping is allowed.

10

WHICH DOOR?

ROB EASTAWAY

You may have heard of the famous American gameshow *Let's Make a Deal*, in which the star prize (a car) was hidden behind one of three doors, and the contestant had to guess which door was the lucky one.

Now there's a new gameshow in town, *Let's Make a BIGGER Deal*, hosted by Tony Macaroni. This time there are five doors instead of three, labelled A, B, C, D and E. A contestant, Kelly, is hoping to win the prize. Kelly is allowed to choose any three of the five doors. If the prize is behind one of those three doors, she wins!

Kelly picks doors A, B and D.

Now – as is always the case on the show – to build up some drama Tony Macaroni opens three doors that he *knows* don't have the prize behind them. On this occasion, Macaroni opens doors A, D and E.

Two doors remain closed: one of Kelly's choices (which is B), and door C.

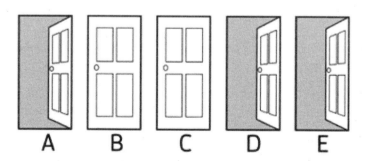

Macaroni says: 'Kelly, do you want to stick with Door B? Or do you want to switch to Door C? You can phone a friend if you want!'

Kelly likes this last suggestion and she rings her friend Jeff. 'Hi Jeff, there are two doors left, B and C. Which one should I choose? Just give me a letter please!'

Unfortunately, Jeff gets flustered under the pressure and just blurts out a letter. What should Kelly do?

Chapter 2

VIRTUAL(LY) REALITY

Quite a few *New Scientist* puzzles have been inspired by situations that have cropped up in everyday life. That includes all the puzzles in this chapter, though some needed to be more heavily adapted and disguised than others. Neither of us owns a yacht in the Mediterranean, for example.

11

BONE IDLE

ROB EASTAWAY

University student Rick Sloth has spent his life avoiding work, and even though it's exam season he still doesn't plan to change his old habits.

He's studying palaeontology, which he thought might be an easy option when he signed up for it, but he's now discovered that it requires rather more study than he was expecting. It turns out there are eighteen topics in the syllabus, and his end-of-year exam will feature eleven essay questions, each on a different topic. Fortunately for Rick, in the exam the candidate is only required to answer four questions in total.

Rick wants to keep his workload to a bare minimum, while still giving himself a chance of getting full marks.

How many topics does he need to study if he is to be certain that he will have at least four questions that he will be able to tackle?

And can you come up with a general formula for how few topics you need to study based on the number of exam questions and the number of topics in the syllabus?

12

SUNDAY DRIVERS

ZOE MENSCH

The single lane road around Lake Pittoresca is stunning if you enjoy scenery, but a pain in the neck if your goal is to get to your destination fast.

Four couples staying at the Hotel Hilberto are planning a day trip to the village of Paradiso at the other end of the lake. The driver for each couple habitually takes life at a different speed. Mr Presto likes to go full throttle at every opportunity in his open-top Porsche. Mme Vivace is not quite such a speedy driver. The Andantes, meanwhile, prefer a leisurely drive, while the inconsiderate Mr and Mrs Lento creep along in second gear, ignoring any honking horns behind them.

Needless to say, if a car finds itself behind a slower car, there is no choice but to follow at the slower speed, forming a larger 'clump' (a clump can be formed of any number of cars, from one upwards).

On Sunday morning after breakfast all four couples set off. As luck would have it, they are the only four cars on the road. By the time they arrive at Paradiso they are in two clumps. Later, the four couples head back in reverse order, and arrive at the hotel in three clumps. Mr Presto looks particularly stressed because he didn't have much opportunity to put his foot down on the journey back.

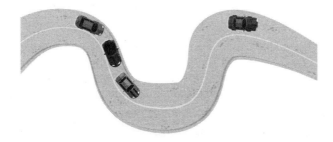

In which order did they set out in the morning?

13

LEAGUE OF NATIONS

ROB EASTAWAY

The TV sporting highlight of my childhood was always the Five Nations Rugby Championship, a series of matches between England, Scotland, Wales, Ireland and France. Every fortnight on Saturday afternoon there would be two matches, with the fifth country having the day off.

The fixture list had an elegant symmetry to it. Each country played every other country once, with two matches at home and two away, and each country alternated between playing at home and away.

I have a hazy memory that one year, the fixtures on the opening Saturday of the Five Nations were Ireland v England (in Dublin) and France v Wales (in Paris). On the third Saturday, Wales played at home.

If my memory is right, what were the final two matches that season?

14

FASTEST FINGERS

HOLLY BIMING

The contestants were lined up, each hoping to get into the Millionaire chair. First they would need to get through the 'fastest fingers' round.

The host cleared his throat:

'List these animals in order of the number of legs they have, starting with the most:'

A Fettlepod	B Eldrobe	C Sentonium	D Quizzlehatch

Guessing blindly, Jasmine went for CDBA, Virat chose CBDA, and Finnbarr picked ADCB, but none got all four right. In fact they all got the same number of answers correct in the right position.

Which has more legs, a Fettlepod or a Sentonium?

15

RESHUFFLING THE CABINET

HOLLY BIMING

The Ruritanian Prime Minister is in a bit of a fix. Thanks to a series of incompetent policy decisions, all five of her senior ministers need to be axed from their current posts. However, the PM cannot afford to sack them completely, because they'll wreak havoc if they are relegated to the back benches.

She has a solution: a reshuffle! She will simply move each of the five ministers to one of the other top posts, but no two of them will directly swap with each other.

Anerdine will move to the department of the person who will become Chancellor. Brinkman will replace the person who will be the new Home Secretary. Crass will take over the post being vacated by the person who will take Eejit's job. Dyer will be appointed as Health Secretary even though he's been lobbying to become Chancellor. The current Defence Secretary will take charge of the department of the person who is becoming the Education Secretary.

Can you figure out who currently has which job, and where they are moving to?

16

AMVERIRIC'S BOAT

ROB EASTAWAY

The well-known billionaire Mr Amveriric keeps a yacht in a private dock in the Mediterranean. It is tethered to the quay by a rope.

Last time his staff tied up the boat, they left too much slack in the rope, so the boat is now one metre away from the quay when the rope is taut. Hearing that a storm is on the way, Amveriric realizes that the boat might get smashed against the wall by the buffeting wind, so he sends his henchman, Benolin Chestikov, to shorten the rope.

Seeing that the boat is one metre from the wall, Benolin decides he will pull the rope horizontally by one metre. As he pulls, the boat moves in horizontally.

The question is: will the boat reach the wall or not? (And can you prove the answer to yourself without resorting to trigonometry?)

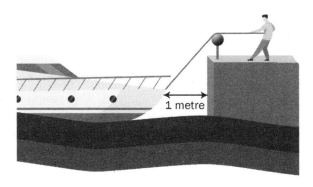

17

EXPRESS COFFEE

DEREK COUZENS

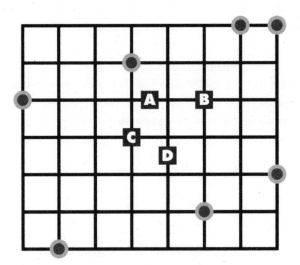

The streets of New Addleton are set out in a rectangular grid.

Seven coffee vendors (indicated by the circles) have stalls at metro stations, and want to set up a central depot from where they can collect supplies each morning. They want to keep their combined cycling distance from stall to depot to a minimum. Pat has picked out four candidates for where to put the depot: A, B, C and D.

'Are you sure one of those four is the optimal location?' asks Shahin. 'I suppose we could check out the total vendor–depot distance for every point on the grid.'

'No need, I can confidently tell you the best place just by looking at the diagram,' announces Kim.

Which location does Kim recommend, and why is she so confident?

18

SEVENTH TIME LUCKY?

HOLLY BIMING

Septa knows that the four digits of the PIN for her bank card are different, but apart from that her mind is a blank. She's had six attempts so far, with no success:

5 7 2 6

7 3 5 8

1 1 9 1

7 6 2 8

4 8 8 2

9 3 0 7

Her bank has a rule of 'seven strikes and you're out', so she has just one more attempt before the machine swallows her card. As it happens, she had exactly one correct digit in the right position in each of her guesses.

What's her number?

19

THE TWO EWES DAY PARADOX

ROB EASTAWAY

Farmer Giles is thrilled that his rare-breed sheep, Lewecy, is pregnant with twins. He's hired an expert from the genetics clinic to find out more about the lambs. The clinic has a reliable new prenatal test that looks for fragments of the lamb's Y-chromosome circulating in Lewecy's blood. The test has come up positive, which means that at least one of the lambs will be male.

'I'm pleased that there'll be another ram on the farm,' thinks Giles, 'but I do hope the second lamb will be female so that I'll have two ewes on the farm next year. There's roughly a 50-50 split between male and female lambs, so if one's a male then it's odds-on that the other lamb will be female.'

Is Giles right to be optimistic?

20

THE HEN PARTY DORM

ZOE MENSCH

Ten friends have rented a dormitory for the night of a hen (or bachelorette) party and each has picked a bed before heading out on the town. At 2 a.m. they get back, a little the worse for wear. Amy, the first to arrive back, can't remember which her chosen bed was, so she just picks one at random. The next one back, Beth, heads for her own bed, but if it's already been taken she randomly picks another. The remaining friends adopt the same approach of going to their bed if available, and randomly picking another if not.

Janice is the last to arrive back. What's the chance that her own bed is still empty? And was Janice more or less likely to find her own bed empty than Iona, who got back just before her?

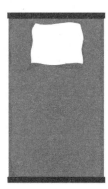

Chapter 3

ALL THINGS CONSIDERED

When was the last time you stopped and thought about your socks? Or saw the missing pieces in your jigsaw puzzle not simply as a frustration but as a frustration *and* an inspiration for a puzzle of a different sort? The head-scratchers in this chapter are inspired by the ordinary and everyday objects of our lives (and cake, which might not be as everyday as we'd like). These puzzles will entertain, challenge, and might just cause you to see the mundane things around you in a new light. That is, as long as you can get the lightbulb turned on.

21

CUTTING THE BATTENBERG

ANDREW JEFFREY

Lady Federica von Battenberg has baked a cake for her daughter Victoria's birthday party. Eight children will be attending in all so eight slices are needed.

She could, of course, make seven vertical cuts to make eight identical slices. But Victoria has heard it's possible to cut the cake into eight identical slices with only *three* straight cuts of the knife.

To be clear, not only must each slice be the same shape, they must all have the same amount of pink and yellow sponge and the same amount of marzipan on the outside. There's more than one way for Lady Federica to achieve this. How many can you find?

22

A JIGSAW PUZZLE

ROB EASTAWAY

Sabrina pulled the old jigsaw box off the shelf, blew off the dust, and gazed at the picture of a thatched cottage on the lid. Under the picture in big bold type was written: '468 pieces'.

Sabrina tipped out the contents. It didn't *look* like 468 pieces. She started to count, but realized this was going to take a long time. What if she just counted the pieces with straight edges? That might be quicker, and if any of those pieces were missing, that would confirm that it wasn't a complete jigsaw.

If only she knew how many edge pieces there were, including corners.

Since there was nothing unusual about the shape of the jigsaw, how many should she expect to find?

23

THE NINE MINUTE EGG

DAVID BEDFORD

I like my eggs to be boiled for exactly nine minutes. The problem is that I have no way to measure time except for two egg-timers: one measures precisely four minutes and the other measures precisely seven. There's more than one way to set up the timers to measure exactly nine minutes, but I'm keen to eat my egg as soon as possible. Can you help?

24

MURPHY'S LAW OF SOCKS

ROB EASTAWAY

I'm convinced that my washing machine eats socks. Every time I wash a load, another sock disappears. Last week I ran out of socks so I bought myself three new pairs.

Every time I wash clothes I just bung in all my socks, including the odd ones. What's the chance that after my first three washes I will be left with three odd socks? Indeed, what's the chance I'll even have one pair intact?

25

REARRANGING BOOKS

ZOE MENSCH

Every week it's Talib's job at the library to put books back in order on the shelf. This week, he finds that the ten-volume encyclopaedia has been completely mixed up. He has to put them back in order, and since the books are heavy he wants to pull as few volumes off the shelf as possible. A move consists of taking a book off the shelf and sliding other books to the side to make space if necessary before putting it back in its new location. What's the smallest number of moves he needs to make in order to rearrange the books back in the order 1 to 10 (left to right)?

26

HIDDEN FACES

STEVE WAIN

While window-shopping at a toy store, my partner and I came across a set of poker dice, which were presented in a semi-transparent plastic case and were partially obscured at the back of the window's display.

My partner took one look and bet me that I couldn't tell them what the sum of all the touching faces would be. I accepted and guessed correctly. Could you?

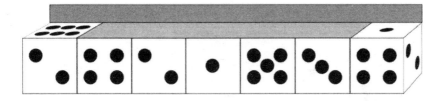

27

BIRTHDAY CANDLES

KATIE STECKLES

It is Mateo's birthday and he has a cake with seven candles on it, arranged in a circle – but they're trick candles. If you blow on a lit candle, it'll go out, but if you blow on an unlit candle it'll relight itself. Since Mateo is only seven his aim isn't brilliant. Any time he blows on a particular candle, the two on either side also get blown on as well. Is there a way he can blow out all the candles? If so, what's the fewest number of puffs he can do it in?

28

LIGHTBULB MOMENT

ROB EASTAWAY

A tall office building is being rewired. There's a staircase, but the lift is out of action.

There are four identical-looking wires, A, B, C and D, feeding into a pipe in the basement ceiling. You are confident that it is those same four wires that emerge from a pipe on the top floor. Unfortunately the wires have become tangled, so you can't tell which wire becomes 1, 2, 3 or 4.

To find out, you can join two wires together in the basement (for example A and C) and you can attach two wires at the other end to a lightbulb and battery (for example 1 and 3). If the bulb lights, you have made a circuit.

Starting on the ground floor, what's the smallest number of lightbulb flashes that you need in order to figure out which wire is which? And how many times do you need to climb the stairs?

BLOXO CUBES

KATIE STECKLES

Chloe and Clive were hard at work packing BLOXO kids blocks in the warehouse. The blocks came in two sizes of pack - yellows, which are a set of four cubes attached together into a 2 by 2 square, and blues, which are a single cube.

With a pile of blocks in front of them, and a box to fit them in, Clive had become despondent. 'This box is big enough to fit 27 cubes in a 3 by 3 by 3 arrangement, and we need to fit in six yellows and three blues, which does add up to 27 cubes. But I'm not sure they're going to fit. These awkward four-packs are going to stick out the top.'

'I'm sure we can work out a way to arrange them so they all go in!' exclaimed Chloe.

'Well, I'm confident they won't fit,' returned Clive, standing firm.

Who is right? Can you prove it?

30

KNIGHT NUMBERS

PETER ROWLETT

My son is obsessed with chess and has been acting out chess moves everywhere we go, running like a bishop and jumping like a knight across tiled floors. He was tickled to notice that on the numberpad of my keyboard he could type the number 27 using a knight's move, because the move from 2 to 7 is an L-shape, like a knight moves on a chess board.

Alas, he can't make 27 using a bishop move. Bishops move diagonally any number of spaces, so a bishop could make a number like 484 or 9157. A knight could make numbers like 167 or 8349.

But yesterday he made a happy discovery: a three-digit knight number that is exactly 27 more than a three-digit bishop number. (Actually, I found I could append the same digit to the front of each of his numbers and still have a knight number that is exactly 27 more than a bishop number.) What numbers did my son find?

Chapter 4

⌐

FIGURING IT OUT

Numbers are considered the universal language, though it sometimes requires a bit of effort to figure out exactly what they're talking about. The puzzles in this chapter are devoted to numbers and calculations, but don't let that worry you. They're designed to be fun and accessible and range from finding scores of football matches to growing food on Mars. Some of these provide surprising insights into the facts and figures around us while others are clever games and diversions that require a bit of mental tinkering to uncover the solution. So grab a pencil and a scratch piece of paper and go figure.

31

SUM THING WRONG

ROB EASTAWAY

Amira presented her homework to her teacher.

'Wrong, Amira. Check your working.'

'I promise you it's right. It's just that I've used a code. Every digit that I've used represents a *different* digit, and the same digit is always represented by the same 'wrong' digit. So for example maybe I replaced all the "6"s with "4"s. Or maybe I did something else...'

'You're giving me a headache, Amira.'

What is the correct sum?

32

CAR CRASH MATHS

BEN SPARKS

Two cars that are the same model, one blue, one yellow, are on a motorway. The blue car is on the inside lane driving at 70 mph, while the yellow one is speeding in the outside lane at 100 mph. At the instant when they are neck and neck, both drivers observe a fallen tree lying across the road some distance ahead. Both immediately brake. The two drivers both apply the same constant braking force. The blue car pulls up centimetres short of the fallen tree. At roughly what speed does the yellow car hit the tree?

(a) 10 mph

(b) 30 mph

(c) 50 mph

(d) 70 mph

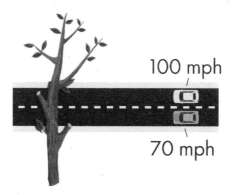

Have a guess. And dredging up your school physics, can you then work it out and see if you're right?

33

SOCCERDOKU

ROB EASTAWAY

'Football league tables are a bit like accounts,' says Harry the book-keeper. 'The debits and credits must balance. For example, victory for one team means defeat for another, so the total games won must be the same as the total games lost. And every goal scored *for* one team is a goal *against* another one.'

Harry's insights will help our league's archivist. The newspaper cutting with the results of the 1993 season is now smudged, and several entries are illegible. The teams played each other once, and this is how the season ended:

	Won	Drawn	Lost	For	Against
United	2	*	0	2	*
Rovers	2	0	*	4	1
Albion	*	*	2	3	3
Town	*	1	2	0	5

Can you fill in the blanks and work out the scores in all the matches?

34

ALL SQUARES

DAVID BODYCOMBE

Here are two puzzles about squaring numbers.

a. I met Natalie the other day. She wasn't prepared to tell me her age, but she did tell me that in the year N^2 she will turn N years old. What year was she born?

b. Can you work out $(68^2 - 32^2)/(59^2 - 41^2)$ without using a calculator? A gold star if you can do it without having to square any of the numbers.

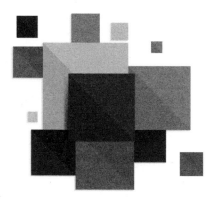

35

SQUAREBOT

CATRIONA AGG

'What's that you're holding, Squarebot?'

'Square.'

I have encountered Squarebot before, and am suspicious.

'Are you sure, Squarebot? I can see it must be a rectangle, because you've drawn it neatly on squared paper. But I can't count the squares without breaking social distance. How wide is it?'

'16.'

'And its height?'

'16.'

'Sounds like a square, then. Just to check: what's its area?'

'289.'

'Hold on, Squarebot, that doesn't work; 289 is 17 squared. You're doing that thing where you round every numerical answer you say to the nearest square number, aren't you? And if the answer isn't a number you just say "Square"?'

'Square,' chuckles Squarebot.

'So the width might actually be 17? Or 18? Or 15? Or even 20?'

'Square,' grins Squarebot.

Can you think of a question that I could ask Squarebot to find out if the rectangle really is a square?

36

CHRISTMAS GIFTS

ZOE MENSCH

a. According to the old Christmas carol, on the twelfth day of Christmas my true love gave me 12 drummers drumming, 11 pipers piping, ten lords a-leaping, nine ladies dancing, eight maids a-milking, seven swans a-swimming, six geese a-laying, five gold rings, four calling birds, three French hens, two turtle doves, and a partridge in a pear tree. So by the twelfth day I had received a total of twelve partridges. But which gifts had I received most of?

b. This happened to me a few years ago, and I decided that I wanted to give these gifts back. Starting on 26th December 2019, I gave one of them to my true love each day. On which date did I give them my final gift?

37

TIGHTWAD'S SAFE

ROB EASTAWAY

Tom Tightwad keeps his money in a safe, the code for which is a ten digit number that uses every digit between 0 and 9.

Fearful of forgetting the code, Tightwad has left it on the front of the safe, but he has disguised it in a special way. He wrote the first five digits along the top of a five by five grid, and the last five digits down the side. Then he multiplied each digit along the top by each digit down the side, filling the grid with 25 numbers. Finally he erased his code number from the top and side so that only the grid remained.

Unfortunately, not only did he erase his code, he also erased most of the numbers in the grid. And of what remains, several digits are so hard to read that they've been replaced with an 'X' (so for example the code's first digit multiplied by its eighth digit is 'twenty-something').

Can you crack the safe?

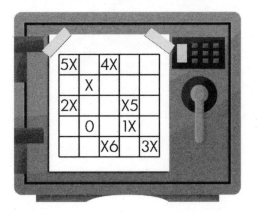

38

THE CARD CONUNDRUM

ZOE MENSCH

Carl scribbled an equation on a scrap of card and left it on a table.

$$\frac{X}{6} = 1 - \frac{(1-X)}{8}$$

Bob found the card and realized that this was just a straightforward algebra problem. 'I've found the solution,' he announced a minute later, dropping the card back on the table and leaving the room.

Alice overheard him, walked over and picked up the card. After a while she announced: 'That's strange, I've found *two* solutions.'

Even stranger, Alice's solutions were different from Bob's.

What were they?

39

MARTIAN FOOD

ROB EASTAWAY

It is the year 2100 and the Mars Pioneers have built an agri-bubble in which they will be able to grow their own food. The crop is a form of grass that grows at a steady rate and can be harvested and turned into nutritious protein snacks (yum!). Now it's time to populate the planet.

The scientists have figured out that if there are 40 adults living in the bubble, the crop will only feed them for 20 days. However with only 20 adults, the crop will keep them going for 60 days - so half as many adults can survive for three times as long! Why? Because without over-grazing, the crop is able to replenish itself!

Of course, the pioneers want a food supply that keeps the population sustained indefinitely. Based on the numbers above, how many people should be in the first Mars cohort?

40

DIFFY

ROB EASTAWAY

Diffy is a children's subtraction game. You choose any four whole numbers between 1 and 12 and place them on the corners of a square (e.g. 3, 1, 9 and 12, going clockwise). Then you find the difference between the numbers at neighbouring corners and write the answer at the midpoint between them (2, 8, 3 and 9 in the example below).

The midpoint numbers are joined to form a diamond, and then the process is repeated until you end with four zeroes in a square (which always happens, eventually).

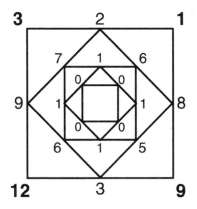

In the example there are five squares altogether, including the zero square at the end. But if you choose the right starting numbers you can get more than five Diffy squares.

Using whole numbers between 1 and 12, how many Diffy squares can you create?

- More than 6 Diffy squares – High Achiever
- 10 Diffy squares – Genius

Chapter 5

A MATTER OF TIME...

We hope you've been enjoying the time you've spent with these puzzles so far, because, one way or another, that's time you can't get back. What we can give you, however, are several of our best puzzles about time. Seconds and minutes, months and years, and clocks of all sorts lend themselves to a variety of tricky challenges, and we've curated some of our favourites in this section. You can test your puzzling speed or go at a leisurely pace, but the clock is ticking, so let's get into the chapter. It's about time.

41

PIECES OF EIGHT

CHRIS TIERNAN

I have a 12-hour digital number display alarm clock. As is normal on digital clocks, each of its four digits is made up from a seven-segment display.

I go to bed when the display is at its dimmest and get up when it's at its brightest.

How long do I spend in bed?

42

WHICH FLIPPING YEAR?

ZOE MENSCH

2019 was an example of a year that can be 'flipped', because on an old-fashioned calculator display it still forms a four-digit number when the calculator is turned upside down ('flipped'):

The difference between a flippable year and its flipped version is called the *flipping difference*, and for 2019 the flipping difference was 6102 - 2019 = 4083.

Since the arrival of the Romans in Britain in 43 C.E., which year has had the biggest flipping difference?

43

SIX WEEKS OF SECONDS

BEN SPARKS

Which number is bigger: the product of all the numbers from 1 to 10 (sometimes written as 10!, or ten factorial), or the number of seconds in six weeks? Can you work it out without resorting to a calculator?

44

LARA'S BIRTHDAY

ANDREW JEFFREY

'It's amazing,' said Lara. 'Today is the 29th of the month and I am 29. Tomorrow is the 30th, and it's my 30th birthday; imagine someone's age matching the date two days running!'

'Not that amazing, is it?' replied Mo, doubtfully. 'Surely that will happen to everyone at least once in their lifetime.'

'It hasn't happened to me,' said Francesca, 'and it never will.'

'Nor me,' said Martha, 'and my birthday's the same week as Francesca's.'

When are Francesca and Martha's birthdays?

45

THE MOUNTAIN PASS

HUGH HUNT

Aaron has spent the night camped at the foot of a mountain, while Bonnie camped at the summit. In the morning, Bonnie sets off down the path to base camp at exactly the same time as Aaron begins his ascent.

At midday they pass each other and nod a greeting, both of them maintaining their constant walking pace. Bonnie gets to the bottom at 4 p.m. and sets up camp, but it isn't until 9 p.m. that Aaron finally reaches the top.

What time did the two hikers set off in the morning?

46

A WELL-TIMED NAP

CHRIS HEALEY

I keep an analogue clock by my bed. One afternoon I went for a nap. When I drifted off, the minute hand was pointing directly at one of the twelve numbers on the clock face, and the number of minutes past the hour was exactly the same as the angle (in degrees) between the hour and minute hands.

Later that day, when I woke up, I noticed the same was true again.

How long had I been asleep?

47

TRIPLET JUMP

ROB EASTAWAY

'My Auntie Connie just had triplets, three boys. And she had them on her birthday!'

'Wow, how old is she?'

'I dunno, but she's, like, *really old.*'

'Do you think the boys' ages will ever catch up to your auntie's? If you add all three together, I mean.'

'I'm not sure they'll ever add up to her age *exactly* – I think it depends on how old she is now.'

Assuming they all live to a ripe old age, what are the chances that there will come a date in the future when the three boys' ages add up exactly to their mother's age? And would it make a difference if the boys weren't born on their mother's birthday? (To be clear: your 'age' is how old you were on your last birthday, so it's always a whole number.)

48

HALF TIME

BRIAN HOBBS

'I couldn't help but notice,' said Sherlock, 'that in the struggle, a clock, with numbers one through twelve, had fallen off the wall.'

'How interesting,' Watson lied.

'In doing so, the clock face had broken into two distinct pieces. The numbers on one of the pieces added up to an odd number, while the numbers on the other piece added up to an even number. All digits were present and accounted for. Now what do you get when you add those two numbers together?'

Watson scoffed. 'Not your best puzzle, Holmes. One plus two plus three, all the way up to twelve, whatever that number is.'

Sherlock smirked and gave me a sideways glance. 'So what do you say, friend? Is our dear Watson correct?'

49

SEEING RED

HUGH HUNT

The traffic lights near me are annoying: green for only ten seconds and red for 90 seconds. I do this ride on my bike every day and I first see the lights as I approach around the bend when I'm 15 seconds away. I get cross if I miss a green light that I could have got through if I'd gone a bit faster. But what makes me *really* cross is if, after a lot of extra effort to get through a green light, it turns red just as I approach. I know I can speed up by about 25% or I can slow down. What should my strategy be if the lights are green when I first see them? And what if they're red? And how often might I be *really* cross?

50

ONE OF THESE DAYS

BRIAN HOBBS

Dear class,

I told you that we would have a quiz next week, and many of you have asked on which day it will be given. An understandable question, so I have helpfully provided the answer below.

A: The quiz will be on Friday, or else Tuesday.

B: The quiz will be on either Monday or Thursday.

C: Either statement B or D is false, but not both.

D: Either the quiz is on Monday or Wednesday, or exactly two of statements A-E are true (or both).

E: Either the quiz is on Tuesday or Thursday, or more than two of statements A-E are false (or both).

Hope that helps.

Your favourite maths teacher,

Mr. Gordon

On which day will the quiz be?*

---

* When Mr Gordon says 'A or B', he means either A or B or both of them – unless he states otherwise.

Chapter 6

MIND GAMES

Are you prepared for a battle of the wits? The puzzles in this chapter are all about competitions and the strategies behind them. Some of these are based on games that are fun and familiar, like carnival games and escape rooms, while others we hope you haven't experienced firsthand, like a three-way gun duel or being hunted by spiders. In any case, it's all in good fun, and we think you'll enjoy working out the optimal strategies and uncovering ways to 'game the system'. May the odds be ever in your favour.

51

CATCH UP 5

ROB EASTAWAY

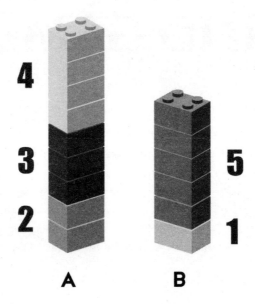

Catch Up 5 is a two player game using five stacks of toy bricks of height 1, 2, 3, 4 and 5. The aim is to end with a taller tower than your opponent. Player **A** starts by taking a single stack of any height. (In the example game shown they chose the '2' stack). **B** then takes as many stacks as they want, stacking them up until their tower is *the same height as or taller than* **A**'s, which ends **B**'s turn. Here, **B** took the '1' stack then the '5'. **A** now does the same, stacking until their tower is at least as tall as **B**'s. In this game **A** took the '3' stack then the '4'. The players take turns until all of the stacks of bricks have been used up, so player **A** won this game.

Imagine you are going first in a game against a *Catch Up 5* expert who always plays the optimal move when it is their turn. Which piece should you choose?

52

TAKING THE BISCUIT

ZOE MENSCH

Alpha and Betty play a rather greedy game with biscuits. There are eight chocolate biscuits in one jar, and four lemon biscuits in the other. Each player can take biscuits in one of two ways.

Either:

- take any number of biscuits from one jar, or

- take an equal number of biscuits from both jars.

The player who takes the last biscuit wins the game and gets to keep *all* the biscuits.

Alpha is set to go first. What biscuit or biscuits should she take?

53

A MEXICAN STANDOFF

ROB EASTAWAY

Three Spaghetti Westerners, by the names of The Good, The Bad and The Bumbling, have decided that the only way to resolve their differences is with a three-way duel, otherwise known as a Mexican standoff.

The three position themselves in a triangle, each armed with a Colt-45 and an unlimited supply of ammunition.

As you might expect from this clichéd scene, Good is a brilliant marksman: he hits his target 99% of the time. Being the villain, Bad is more hit-and-miss: his success rate is 66%. And Bumbling, in his role as the comedy relief, only hits the mark 33% of the time.

On a count of three, each will draw their gun and, using the best strategy they can, will keep shooting until they are the last man standing.

What (roughly) is the chance that Bumbling will survive?

54

MY FAIR LADYBUG

BRIAN HOBBS

'Ooh, I want that one,' says my young daughter, pointing to the stuffed ladybug cushion on display. She loves anything to do with ladybugs (in your country perhaps you call them ladybirds). 'And it's only five dollars per guess!'

We're standing at Rip-off Rick's Cup Shuffle carnival game at the local fair. In the game, he hides a bean under one of four cups and shuffles them before placing them in a line: A, B, C and D. Provided you're loaded with cash, you can then buy as many guesses as you want to locate the bean.

The catch is, between each guess, Rick waves his hand over the cups, and the bean 'magically' moves to another cup. My buddy worked here last summer and let me in on a little secret: Rick can only move the bean to an adjacent cup (e.g. from B to either A or C).

I've seen these cushions in the shops and I know I can get them for $16, so no way am I spending more than $20 on this game. Since each guess is 'only' five dollars, can I guarantee a ladybug in four guesses?

55

DUNGEONS AND DIAGRAMS

BRIAN HOBBS

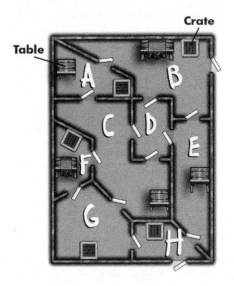

'How has somebody managed to leave behind their shoe?' grumbles Bernie the intern as he picks up the lost property after the latest team has left the Incredible Dungeon Escape Adventure (IDEA™).

Bernie's job is to set up the dungeon between teams. Starting in one of the rooms, he walks through every door exactly once, including the two external doors and the secret door that's located in the wall directly behind one of the crates. He locks each door behind him and puts back all the furniture the previous team needlessly flipped over. Finally, he puts the Exciting Treasure Chest (ETC™)next to the table in the final room before leaving by the hidden staff-only trap-door exit in the floor to take his 7-minute lunch break.

The contestants' map doesn't show the hidden door or the trap-door exit.

Where did Bernie start, where is the treasure, and how do you know?

56

ANT ON A TETRAHEDRON

ALEX MAYALL & ALARIC STEPHEN

Three short-sighted spiders are clustered at the vertex of a wire frame in the shape of a tetrahedron. The spiders know that there is an ant walking around the frame but they have no idea where it is. They'll only be able to spot it when they are practically on top of it. The ant, on the other hand, has brilliant eyesight and can plan its route accordingly to avoid the spiders.

Given that the ant walks slightly slower than the spiders, is there a way for the ant to escape the spiders indefinitely (with a bit of luck)? Or can the spiders find a strategy to be certain of catching the ant?

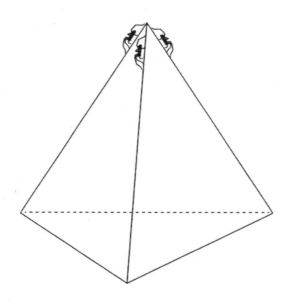

57

CHOPPING BOARD

PETER WINKLER

The Board of Overseers of the Bottlecap Preservation Society has grown to an unwieldy ten members, and so the members have agreed to the following procedure. The board will take a series of votes on whether to reduce its size. A majority of 'ayes' results in the immediate ejection of the newest board member. Then another vote is taken, and so on. If at any point at least half of the surviving members vote nay, the session ends and the board is fixed at that level.

Of course, each member's highest priority is to remain on the board, but aside from that, everyone agrees that the smaller the board, the better.

If logic prevails, how big will the board be when the voting comes to an end?

58

ÉCLAIR-VOYANCE

BRIAN HOBBS

Tom and Amy are colleagues who are both excellent logicians; they speak honestly and accurately, and no bit of good deduction ever slips past them. Another thing they won't let slip past them is the last éclair sitting on the tray at the annual Puzzlers' party. Since neither is willing to back down, they propose a solution. They will take the diamonds out of a deck of cards and remove the ace and face cards, leaving them with the 2, 3, 4, 5, 6, 7, 8, 9 and 10 of diamonds. They will shuffle those together and each draw a card. Whoever has the higher card gets the éclair.

They each take a card and look at it, being careful not to show it to the other.

'Well, I don't know whether I'll win,' says Tom, 'but I hope I do.'

'Same goes for me,' replies Amy. 'Do you know who won yet?'

'No,' says Tom.

'Me neither,' says Amy.

At that, Tom sighs and tosses his card back into the deck. 'You win.'

What cards did they each have?

59

THE GOBLIN GAME

ZOE MENSCH

Annie and Beth are about to play 'Goblin'. Like snakes and ladders, it is played on a 10 by 10 square numbered from 1 to 100, with players starting with their counter off the board (next to square 1) and rolling a single dice, the aim being to be the first player to get to square 100.

However, instead of snakes or ladders there is only one hazard – a Goblin. Each player gets one Goblin, and is allowed to place it on any square they want (apart from square 100) before the game starts. If you land on your opponent's Goblin you lose, and similarly, your opponent loses if they land on your Goblin. If a player lands on their own Goblin, they are safe. If neither player lands on a Goblin, the first to get to square 100 wins (an exact final roll is not required, just getting up to the 100 square is enough).

Before they start, the two players place their Goblins. Annie, who has never played before, picks square 31 because that's her lucky number. Where should Beth place her Goblin to have the maximum chance of winning?

100	99	98	97	96	95	94	93	92	91
81	82	83	84	85	86	87	88	89	90
80	79	78	77	76	75	74	73	72	71
61	62	63	64	65	66	67	68	69	70
60	59	58	57	56	55	54	53	52	51
41	42	43	44	45	46	47	48	49	50
40	39	38	37	36	35	34	33	32	
21	22	23	24	25	26	27	28	29	30
20	19	18	17	16	15	14	13	12	11
1	2	3	4	5	6	7	8	9	10

60

WEATHER OR NOT

BRIAN HOBBS

At the Shady Hills retirement home, Joe and Eileen are arguing over their predictions for the weather.

'I'm telling you, Eileen, it's going to rain tomorrow and the next day, and then it won't rain on the third day.'

'Balderdash. The way my knee is acting up, I'm sure that it won't rain tomorrow, but it will rain for the following two days. I'll bet my Monday night cheesecake on it.'

'I'll take that bet. But what if neither of us is right?'

'Then we'll just keep waiting until there are three consecutive days that match one of our predictions. So if it rains for two days and then doesn't rain, you will win Joe.'

'Deal.'

Two of the staff overhear the conversation. 'This time of year, there's about a 50-50 chance that it will rain on a given day,' says one, 'so I guess their chances are even.'

'You're right about the rain,' says the other. 'But even so, I think one of them has a much better chance of winning.'

Which one, and why?

Chapter 7

ECCENTRIC TALES

Puzzle World is a curious place where all sorts of unlikely coincidental things take place in everyday life. But some of the best puzzles don't pretend to be part of the real world. It's a fantastical world of princesses, shipwrecked sailors, adventurers and old-fashioned eccentrics. This chapter is dedicated to these weird and wonderful characters.

61

THE CAKE AND THE CANDLES

DAVID BEDFORD

Lady Frederica von Battenberg has baked another long, thin rectangular cake for her daughter Victoria (she has a history of doing this; see Puzzle 21). This time she has picked two random points on the top of the cake on which she has placed two candles.

She hands the cake knife to Victoria, who now proceeds to pick a random point along the length of the cake, and cuts across the cake at that point.

Now that the cake has been cut in two, what is the chance that both pieces of cake have a candle on them?

62

VIVE LA DIFFÉRENCE

CHRISTIAN LAWSON-PERFECT

The eccentric manager of the *Bistro Vive la Différence* has a bizarre method of offering discounts to his customers. Each seat has a number, and each customer gets a discount (in euro) equal to the difference between the numbers of the seats either side of them. For example, a guest seated between seats 6 and 1 would get a discount of 5 euro on their meal.

Seven friends have booked a meal, and arrive to find chairs numbered 1 to 7 in order around the table. They figure that with the seats arranged in this order, they can get a combined discount of 20 euro. But they reckon they can do better. They want to rearrange the chairs to maximize their discount, but only have time to make one swap before the waiter comes to take their order. Which chairs should they swap?

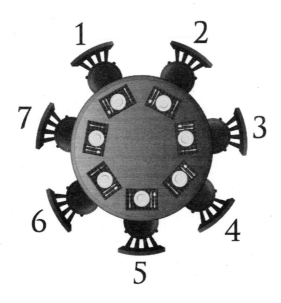

63

DIAMONDS ARE FOREVER

ZOE MENSCH

'Once upon a time,' began Ivan the storyteller, with children at his feet, 'there lived a queen called Factoria who had six daughters. Now the queen had many palaces. In each palace she kept as many crystal vases as she had palaces, and in each vase were as many diamonds as there were vases in that palace. Then one day the queen died, leaving a will:

I leave one vase of diamonds to my loyal servant Fidelio. The rest of the diamonds I will share equally between my daughters. Any remaining, Fidelio will put in this box.''

Ivan reached into his pocket and pulled out a small wooden casket. 'And this is the box!'

'How many diamonds are there?' screamed the children.

'If you can tell me, I will give you the box,' said Ivan.

'But you haven't told us how many palaces...' they cried.

Ivan winked.

Can you, dear reader, figure out how many diamonds were in the box? And how can you be certain?

64

CHANGING THE GUARD

ROB EASTAWAY

Fifteen members of the King's 99th Dragoons are standing on parade.

'Right Turn!' screams the sergeant, and each soldier makes a quarter turn. Unfortunately many of the squad struggle to know their left from their right, and after this manoeuvre, five of the soldiers end up facing leftwards, including Private Perkins, who is in the middle of the row.

Any soldier that ends up face to face with another soldier, knowing that something has gone wrong, now does a 180 degree about turn, and this awkward ritual continues until no soldier can see another soldier's face.

In which direction does Perkins end up facing?

65

YAM TOMORROW

CHRIS MASLANKA

Three shipwrecked sailors discover a crate of yams on the beach. The crate is labelled '100 Yams' but they notice it has been prized open and some of the yams have been pinched, possibly by the monkey they spot nearby.

In the night, Abel wakes and decides he will take one third of the yams - but he can only take a whole number of yams if he first gives one yam to the monkey. Later, Babel has the same idea, but again to take one third in whole yams, he has to first give the monkey a yam; and later still the same thing happens with Cabel. In the morning the three sailors, who have all hidden their secret stash, share out what yams remain equally among them, and this time around the poor monkey receives nothing. How many yams did they each end up with?

66

BLURRI-NESS

HOWARD WILLIAMS

CLICK!

The camera shutter opened and closed just as the creature's head ducked back beneath the surface of the lake, creating a large ripple.

'I got it! I finally got a picture of the Loch Ness monster!' Lily exclaimed, looking at the result on the digital camera. 'It's – blurry again!' She hung her head in defeat.

The boat passed directly over the spot where they saw the creature, but it was nowhere to be found. 'How far away do you think it was when I took the picture?' Lily asked Amelia later that day.

'Well, we were travelling in a straight line at an even two metres per second, and it took us five seconds to reach that large ripple it created, and another ten seconds to get to the other side of the ripple. So that means it was five plus an additional five seconds to the middle of the ripple, which would make it... twenty metres!'

'Sadly, I think your maths is as fuzzy as my photo,' says Lily.

How far away was the monster at the time of the picture?

67

NO TIME TO TRY

KATIE STECKLES

James Blond edges along the corridors of the supervillain's base, and comes to two locked doors, each with a keypad that requires a four-digit code. He'll need to get through one of the doors, but there's no time to guess a four-digit code - the number of possible combinations is staggering!

But wait! Some of the buttons on the keypads are visibly worn down, while others look like they've never been pressed.

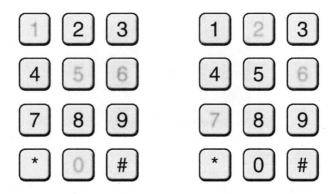

One door has a keypad with four worn buttons, the other has three. Blond only has time to try one door, and he'll have to try all the possible combinations by brute force.

But which of the two keypads will give him fewer combinations to try - the one with four worn buttons, or the one with three?

68

PAINTINGS BY NUMBERS

HOLLY BIMING

When the famous artist Pablo Minestrone held his final exhibition at the Galleria de Cannelloni, he wanted the public to experience his works in the order in which he had painted them. Paintings from his early Green period were to go in Room 1. From there, visitors should move to an adjacent room to see his Mauve works in Room 2. And from there, to adjacent Rooms 3, 4, 5, etc. until they reached the Black paintings (generally viewed as Minestrone's darkest period) in Room 9.

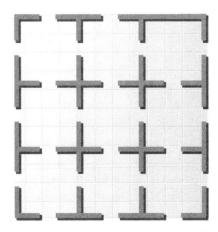

Alas, no details remain of which room was where in Pablo's exhibition. But his widow Bella does recall a curiosity about the numbering of the different rooms: the three-digit number formed by the top row added to the number formed by the middle row equalled the number formed by the bottom row.

Can you recreate Minestrone's Gallery tour?

69

A PIAZZA OF DOMINOES

ZOE MENSCH

Some town squares are designed as giant chess boards, but urban planner Dominica decided to pave her town's new piazza with giant dominoes instead.

Picking different dominoes at random from a set, she laid them down to form a 7 by 7 square, leaving a small square in the centre for a fountain.

Can you draw the outlines of the dominoes that Dominica used in her original design, and figure out which dominoes she left out? (Remember that a full set of 28 dominoes contains every pair of numbers from 0-0 to 6-6. There were no duplicates).

3	0	5	5	1	1	1
2	4	6	2	6	3	0
1	4	0	2	5	4	1
0	4	6	🎁	1	0	2
2	3	3	5	0	3	5
0	0	6	2	1	3	5
5	6	1	2	6	6	3

70

TROUBLE BREWING

HOLLY BIMING

Our office vending machine allows me to get any combination of tea, coffee, chocolate, milk and sugar, each with its own button.

Or at least it did, until accident-prone engineer Bob Klutz gave it the annual 'maintenance' overhaul. Now it has developed a glitch. What happens now is that instead of delivering its own ingredient, each button delivers two other ingredients instead. Each button delivers a different pair, and each of the ingredients is delivered by two of the buttons. However, if two buttons demand the same ingredient, such as tea, they cancel each other out and I don't get tea at all.

The result? If I ask for chocolate with milk, I get tea with sugar. If I ask for tea with milk and sugar, I get those three ingredients plus coffee!

What do I get if I press the coffee and milk buttons?

PART 2

SOLUTIONS, BACK-STORIES AND COMMENTARY

1. CREATIVE ADDITION

The fact that the columns differ by 3 tempts many solvers to just remove the 3, only to realize that the 3 is in the 'wrong' column! Instead of changing the 24 to 21, removing the 3 changes the 21 to 18. There then typically follow attempts to move a card from one column to the other, but with no success. The reason why this is never going to work can be proved mathematically. The two columns add to 45, an odd number, so there is no arrangement of the numbers 1 to 9 in which the columns can add to the same total (they would both need to add to half of 45, which is 22.5!)

This means some lateral thinking is going to be required.

The classic 'aha' solution is to turn the 9 upside down to make 6, so both columns add to 21. But how about:

- Place the 1 on top of the 5, so the 5 is hidden and both columns add to 20. (This was my daughter's solution!) Or place the 3 on the 9 (18), or the 5 on the 7 (19).

- Turn the 5 over to reveal its blank side and place it on the 2 to hide both numbers (columns add to 19). Or the 4 on the 1 (20), or the 6 on the 3 (18).

- Put the 3 to the top right of the 2 to make it 2^3 (= 8) so the columns are both 24. Similarly, place the 1 below and to the left of the 5 (1^5 = 1), making both columns equal to 20.

More controversial solutions submitted by readers included:

- Pour paint over all but 1, 2 and 3, and then move 3 to the other column.

- Place the 3 just beneath the 9 so that it represents 9/3 = 3, so that both columns add to 18.

- Write over the 5 with a thick black felt tip that bleeds through the paper, turn it over and the 5 now resembles a 2. (With the aid of a black felt tip you can probably do several more 'cheaty' solutions).

- Rotate the 8 so that it still looks like an 8. It has 'moved' and the totals are the same as they were before.

- Tear the 8 in half horizontally and put the halves on the right to make 4^0 and 9^0, which both equal 1. (If you want to know why any number to the power of zero is equal to one, there are many excellent explanations online.)

- Tear the serifs (the small dashes at the top and bottom) off the 1 and use the torn pieces to make an exclamation mark (a factorial symbol). Put it next to 3 to make 3! = 3 × 2 × 1 = 6.

- Rotate the 8 by 90° and place it below and between both columns so it contributes to both columns, making them sum to infinity.

- Push the 4 to the left, sweeping away the 1 with it.

- White out the descending bar and one of the horizontal bars of the 7, transforming it into a negative symbol, and place that next to 5, so both columns add to 14.

- Define the addition as being modulo 2 (the remainder after dividing the result by 2) and remove the '1' card.

Are these fair creative solutions, or meh cheating solutions? The answer is... entirely subjective. Most, but not all, solvers regard the solution of turning the 9 upside down as fair, even if it's accompanied by a cry of 'Doh!' Interestingly, however, the answer that was suggested most often by readers was placing the 1 card on top of the 5. The other solutions get a more mixed reaction.

This puzzle provoked a lot of reader comment.

- 'All three of my solutions feel like cheating, though maybe I wouldn't have thought of them as cheating if that word hadn't been mentioned.'

- 'I draw the line of "cheating" at turning things around to change the number.'

- 'The setter said there should be some element of creativity, but this is outright cheating!'

Which just goes to show how blurred the boundary is between creativity and cheating.

Part of the issue with puzzles like this is that there are certain conventions in puzzle setting and solving which are assumed but not stated. When the instruction says you can 'move' only one card, that's generally taken to mean 'move to the other column' (a bit like moving a chess piece). Typically, the more a solution is breaking the silent contract between setter and solver, the more it is deemed to be cheating.

We have put this puzzle first in the book because it lends itself to so much interesting investigation and discussion. However it also risks raising an element of suspicion in the reader. 'If we're expected to be OK with "cheat" solutions here, maybe there will be cheat solutions elsewhere too.'

So let it be noted now that this puzzle is the exception. There are no other puzzles in this collection that involve any unfair trickery, or cheesy interpretations of words.

2. THE H COINS PROBLEM

Inside the box

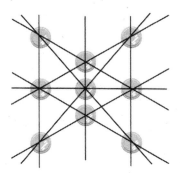

Outside the box

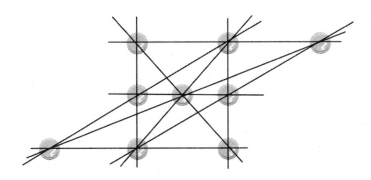

There is a third solution, overleaf, which is just a reflection of the second solution.

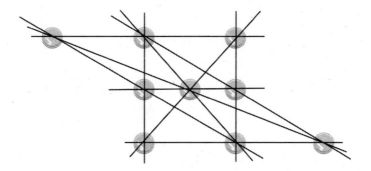

This is the sort of puzzle that after half an hour of head-scratching can leave some people convinced there is no legitimate solution. And even if they find one solution, it's rare that they find the other without a strong hint. The 'trap' that most people fall into is to only try lines that are vertical, horizontal or at 45 degrees.

The puzzle is reminiscent of the classic nine dots puzzle, made famous over 100 years ago by the prolific American puzzle setter Sam Loyd, which challenges you to pass a straight line through each of the nine dots in this square using only four straight lines and with your pencil never leaving the paper:

The trap that solvers fall into if they've never encountered this before is to keep all the lines inside the grid, on the self-imposed assumption that they are not allowed to go outside the box. (The solution is overleaf).

Mind traps like these are common, but hard to overcome even if you've been caught out by similar ones before.

Solution to the classic nine dots puzzle.

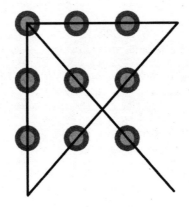

PART 2 ▪ Solutions, Back-Stories and Commentary

3. THE BOOK OF NUMBERS

The first number in the book is eight, and the last is zero. In a dictionary, the convention is that a space is earlier in the alphabet than a letter, so eight hundred would come before eighteen.

The second number in the book is 8,000,000,000 (eight billion) and the second-last is: 2,000,000,002,202 (two trillion two thousand two hundred and two). It's surprising that in a list with an infinite number of members it is possible to state the first two and the last two entries.

If you extend the naming of numbers to everything currently given a name (quintillion all the way up even to googol), then the first entry, second entry, and last entry all stay the same. The second-to-last entry, however, changes to 'two vigintillion two undecillion two trillion two thousand two hundred and two', a vigintillion being the twentieth in the 'illion' sequence, meaning it is a 1 followed by 63 zeroes.

This can be explored much further, of course. What are the third, fourth and fifth numbers? What are the last two numbers in Swedish? There's no end. (We are told that the answer to the Swedish question is that the second-last and last numbers are 8802 and 8808, written as åtta tusen åtta hundra två and åtta tusen åtta hundra åtta. The letter å appears after z in the Swedish alphabet.)

4. LATE FOR THE GATE

You will get to the gate faster if you tie your laces on the travelator, or if you run on the carpet. This is far from obvious. You can prove it with algebra, but there are intuitive explanations. Imagine twins T1 and T2 walking side by side on a carpet leading up to the travelator. T1 stops to tie their laces just before the travelator while T2 stops just on the travelator, a fraction of a second later. T1 will not catch up with T2.

The situation with running is harder to picture, but think about the lead that is gained by T1 running on the carpet. The lead is extended when they walk on the travelator, and T2 can never make up all of the extra lead that T1 built. Algebraically, when T1 runs on the carpet they build a lead of D. They increase this lead by amount X when on the travelator, until T2 is also on it, at which point T2 starts running. T2 closes the gap by D, but never eliminates the lead of X. (It gets more complicated if T1 steps off the travelator before T2 stops running, but the answer still holds.)

As a rule of thumb, if your new action (e.g. tying laces) is slower than your walking speed, do it on the travelator, but if it's faster than your walking speed, do it on the carpet.

This is a rare example of a mathematical problem that found its way into public debate. Because the situation is a familiar one and easily stated, everyone can have a view. However, not everyone is prepared to engage with its mathematical side. Most people think of it in purely practical terms: 'I would just go to the side and tie it, so that other people don't bump into me.'

Even those who think of it purely as a mathematical problem often start with the wrong intuition. When audiences of mathematicians are asked to give their first hunch for the solution to the lace tying problem, they are usually split fairly evenly between four choices.

Typical reasoning goes like this:

- Tie on the carpet – that way you spend the whole time on the travelator going at maximum speed.

- Tie on the travelator – so that you are always moving even when tying your laces.

- It doesn't matter: the distance, speeds and tying times are all fixed, so in the end the result will be the same.

- It depends...on the relative length of the travelator and the carpet, your relative walking speed compared to the travelator, etc.

There is one added twist. What if the travelator is moving backwards? Plenty of people will admit to having experimented with walking up a down escalator, especially after a few drinks. It turns out that if the travelator is moving backwards, you are better off tying your laces on the carpet. And if the travelator is moving backwards at least as fast as you walk it doesn't matter where you tie your laces, because you'll never get to the far end in any case.

5. BUS CHANGE

The easiest way to work out the total is to start from the largest coins and work your way down. You can't have any £2 or £1 coins, but you can have (only) one 50p coin. Then you can add up to four 20p coins, and still not be able to make £1. From here, you can't add in any 10ps (or else you'd have 50p + 20p + 20p + 10p = £1) but you can play the same trick again with 5ps and 2ps, adding in one 5p and four 2ps. This gives a total of £1.43.

This puzzle can be applied to any currency, though the totals are different. In the USA, for example, where the coins are 1, 5, 10 and 25, the most you can have without being able to pay exactly a dollar is a mere 119 cents. The general tactic to figure out the answer in any currency is to work down from the largest coin to the smallest and take as many as you can of each. Many countries have eliminated coins worth less than 5 cents/pence etc, for the simple reason that those coins are barely worth the metal they are minted on. The UK has nostalgically held onto 1p and 2p despite their rapidly diminishing usefulness. On the positive side, it has led to this delightful puzzle.

6. DARTS CHALLENGE

The lowest score you can't get with three darts is 163. All other scores below 163 are achievable in three, for example 161 is 60 + 50 (bullseye) + treble 17. With two darts, 103 is the lowest score that you can't get, and with a single dart it's 23. (Given that you have to finish a game of darts on a double or with a bullseye, the lowest score from which you can't *finish* with three darts is 159.)

Darts scoring is one of the last bastions of mental arithmetic, and ironically the worse you are at darts, the better you get at calculating ('nineteen plus treble five plus eleven makes...?'). The numbering of the traditional dartboard is cleverly designed to penalize the selfish player. Tempting as it is to aim for treble 20 each time (as the professionals do), a small error in direction will see you scoring a 1 or a 5 instead.

Mathematician Zoe Griffiths devised a computer simulation to work out the best place to aim on a dartboard if you want to maximize your points. The model assumed that when you aim at the board, there is random variation in your direction. A top player might have a typical deviation of only a few millimetres, whereas a beginner might only expect their accuracy to be within a few inches.

The results of Zoe's simulation are fascinating. They suggest the following:

Top player – always aim for treble 20

Very good player – aim for treble 19

Decent player – aim for treble 16

Poor player – aim for treble 11 (that's the quarter segment of the board that has the highest average score).

Hopeless player – aim for bullseye, as that will maximize your chances of hitting the board.

7. EVENING OUT

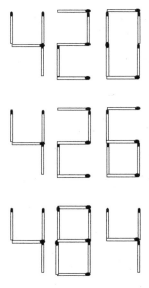

Matchstick puzzles using regular numerals aren't as well known as those involving Roman numerals.

A Roman number matchstick puzzle with a twist is this one.

Make the following addition correct without moving any matches.

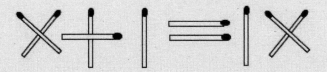

Answer at the bottom of the page.

7: Look at the equation upside down to make XI = I + X

8. THE CAESAR CIPHER

The Caesar cipher was the method that Julius Caesar used to encrypt messages, by shifting letters by a fixed number of places up the alphabet. If you shift the word THREE by four places you get XLVII, which is 47 in Roman numerals. Coincidentally, if you convert the letters of the alphabet into numbers, A = 1, B = 2, then C = 3, and C + A + E + S + A + R = 47.

This is a neat addition to a collection of Roman numeral puzzles and curiosities. When Angus Walker first posed the puzzle, he only knew of the first answer. However when it was posed to a group of maths teachers, several of them quickly found the letter-numbering solution, a surprising coincidence that added to the intrigue of the puzzle.

A puzzle that requires a combination of numeracy and general knowledge is this:

In which historic location in London would you find the seven Roman numerals M D C L X V I in that order?

Answer at the bottom of the page.

8: MDCLXVI represents the year 1666, and this date can be found inscribed on the Monument to the Great Fire of London, which is at the North end of London Bridge.

9. SYMMETRIC-L

Here are the two symmetrical ways of putting the three Ls together. The first is a little like a fish, the second resembles a heart.

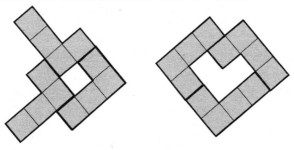

This piecing-together puzzle only involves three pieces yet most people find it deceptively difficult to find even one solution. Why is this? Each L is made up of four unit squares. In how many distinct ways can you put together two of these L-pieces? Let's suppose you limit yourself to those situations where you could glue the two pieces together, and where they have to join along a whole number of units. For example, this pairing touches along a length of one unit:

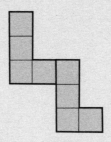

And this is glued along two units:

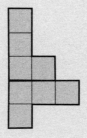

Given that L-pieces can be placed either way up, we reckon there are over 100 distinct ways in which you can join two L-pieces together (by 'distinct' we mean that you couldn't place one pair on top of another even if you flipped it over).

Not many of these pairings are symmetrical, but they might still end up being part of the symmetrical three piece solution, because the third piece might restore symmetry (as is the case in the 'fish' solution above). For each pair of pieces, there are typically over 100 ways of gluing on a third piece. $100 \times 100 = 10,000$, which means that there are over 10,000 possible combinations to explore (we haven't counted them all). So without a bit of smart short-cutting, no wonder this puzzle takes a very long time to solve.

10. WHICH DOOR?

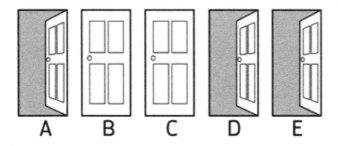

A   B   C   D   E

It doesn't matter which letter Jeff suggests. If he hadn't panicked, he could have asked Kelly which of the doors she had initially chosen. As it stands he has no information to differentiate between the two doors, so from his point of view the prize is equally likely to be behind door B or door C. But to maximize her chance of winning, Kelly should stick with door B regardless of what Jeff says. (In fact, the option of phoning a friend is a bit of a red herring in this puzzle.)

There is a 60% chance that the prize is behind door B, and only a 40% chance it is behind door C.

Why? Kelly had a 60% chance of picking the prize when she chose three doors out of the five. Macaroni then opened three doors that he knew had no prize, so he has revealed no new information, and the probability has therefore not changed.

It's not often that puzzles generate angry correspondence, but this one certainly did. We'll get to the letters shortly. First, some background.

This puzzle is a variant of the so-called Monty Hall problem, which has become a classic in puzzle circles. The original is based on a 1970s TV gameshow called *Let's Make a Deal*, in which the star prize at the end of the show was hidden behind one of three doors, call them A, B and C. The contestant would pick a door (A, for example), and the show's host Monty Hall would then (according to legend at least) open one of the other two doors that he knew didn't have the prize behind it, and offer the contestant the option of either sticking with their original choice or swapping to the other closed door.

When people first encounter this problem, their hunch is usually to reason that since the prize could be behind either of the remaining doors, the chance of it being behind door A is now 50%, so they might as well stick. The reality, however, is that so long as the rules have been stated clearly as above, the contestant should swap doors. The chance that the prize was behind Door A at the start was 1/3, and since the host was always going to be able to open a door that had nothing behind it, the chance that Door A is the winner is still 1/3 after one door is opened. The chance that it's behind the other closed door is now 2/3.

The Monty Hall problem was first posed as a reader question in *Parade* magazine's 'Ask Marilyn' column. The columnist Marilyn vos Savant's answer, that the contestant should swap doors, generated thousands of angry letters from readers who strongly disagreed, many of them from academic mathematicians. It was several years before the controversy subsided, but thanks in part to a discussion of the puzzle in Mark Haddon's bestselling book *The Curious Incident of the Dog in the Night-Time*, it is now quite widely known that your best strategy is to swap doors.

For the *New Scientist* variant, the setter sneakily changed the scenario to selecting three doors out of five doors so that the best strategy would now be to stick and not to swap. It was a deliberate twist to catch out those who assumed that in these puzzles you should always swap.

Angry emails flooded in. Here are three of the many:

'The solution is incorrect. Kelly increases her chance of winning to 4/5. Lesson? Ring the right friend for help.'

'The solution is wrong on various counts. The probability that the prize is behind the other door is 2/3.'

'At the risk of inviting the torrent of mockery visited upon Marilyn vos Savant, I have to suggest that Rob Eastaway's solution is not correct, Jeff should have told Kelly to change her choice.'

The level of correspondence was high enough that the *New Scientist* editor considered publishing a clarification the week after the solution had been published, but in the end it was decided to close the matter. Some people will never be persuaded.

Chapter 2 **VIRTUAL(LY) REALITY**

11. BONE IDLE

Rich needs to study eleven topics in order to guarantee four questions will come up in the exam.

To figure out how many topics you need to study, imagine the worst case scenario. In this palaeontology exam, the maximum number of topics that you could study, only to find that none of them come up in the exam is seven (= 18 – 11). So if Rick studies seven subjects, plus an additional four - i.e. eleven topics in total - then he is guaranteed to have at least four questions to choose from in the exam. In general, if there are T topics, E exam questions and Q questions to be answered, then to be certain that at least Q of your topics will come up, the minimum number of topics S that a lazy student should study is given by $S = T - E + Q$. So with 20 topics, of which you need to answer 5 out of 10 questions in the exam, the formula says you need to study at least 15 topics.

This puzzle was inspired by an enquiry sent in by a listener to BBC Radio 4's *More or Less*, who appeared to be speaking from experience. We changed the subject being studied to palaeontology to give us an excuse for the punny title.

Question spotting and selective study has no doubt been a strategy of students for centuries, though the assumptions behind the above formula are risky. Just because a topic comes up in the exam doesn't mean it'll be a question that you like. And sometimes questions combine topics, leaving the work-shy student only partly equipped to tackle them.

But for those who enjoy the brinkmanship of studying the bare minimum needed to be able to tackle an exam, there is a 'Lazy Student Formula' that should give you at least a 90% chance of enough questions coming up in the exam, so long as the exam requires at least three questions to be answered.

The formula is:

$$S = \frac{T \times Q}{E} + 1$$

This isn't based on some fundamental mathematical theorem; it is simply the result of some experimentation to find a simple formula that gives roughly the right results. In our example of the palaeontology exam with eighteen topics, and four out of eleven questions needing to be answered in the exam, the Lazy Student Formula suggests you could get away with studying only eight topics. Good luck!

12. SUNDAY DRIVERS
The cars left the hotel in the order: Presto, Lento, Andante, Vivace

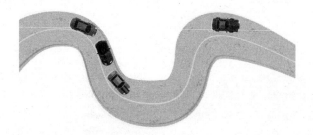

Label the four cars 1 to 4 in decreasing order of speed. Think about car 4, the Lentos. It can't set off first or last as this would lead to a clump of size one (either out or back). The Lentos also cannot set off third as this would give two clumps on the way back. And car 1 (the Prestos) cannot set off last as they would have a clear run on the return, which we are told did not happen.

This leaves four possible orderings on the outward journey: 1432, 1423, 2413 and 3412. Of these, only 1432 would involve two clumps going out and three coming back. So the cars left the hotel in the order: Presto, Lento, Andante, Vivace.

Clumping can be a significant factor in traffic congestion. Traffic engineers – and mathematicians – have devoted significant time to trying to understand how it works so that they can reduce its impact. In the USA, more than half of the states have a law that forbids cars that are moving slower than the 'normal' speed of the other traffic from being in the fast (left-hand) lane, even if the 'normal' speed is above the speed limit. This should in theory remove all clumping, so long as there are at least two lanes on the carriageway. Without such interventions, smarter techniques are needed. A mathematician at MIT found that clumping can be reduced (and traffic flow faster overall) if each driver attempts to keep an equal distance between the car in front of and behind them.

In 2018, Reza Zadeh revealed on Twitter that in a line of N cars on a single lane road, with drivers who have a random assortment of preferred driving speeds, the expected number of clumps can be found by the simple formula: $1 + 1/2 + 1/3 + ... + 1/N$. So, for example, if there is one car there is one clump (of course), with two cars there's an average of 1.5 clumps – and by the time you get to ten cars, you can expect roughly three clumps (and a few very frustrated drivers).

13. LEAGUE OF NATIONS

The last two matches were England v Scotland and Wales v Ireland.

The first matches were Ireland v England and France v Wales. Scotland therefore played every round after that. Wales were at home for the third set of matches, so since home/away alternated, Wales missed the second round of matches, meaning the away teams in round 2 were Ireland and France. Scotland and England must have been the home teams, so Scotland played Ireland and England played France. Following through, we get:

Round 3: Wales v England and France v Scotland

Round 4: Ireland v France and Scotland v Wales

Round 5: England v Scotland and Wales v Ireland

Puzzles based on league tables have been a popular staple for many years (see also Puzzle 33, Soccerdoku, for example). But this particular puzzle had added appeal because it was based on the real fixtures in 1975. Add to that the fact that there appears at first glance to be insufficient information for a unique solution. Several readers got in touch to say that this was one of their favourite *New Scientist* puzzles, partly because those readers were rugby fans, but also because they appreciated that the problem arose from a genuine piece of sport history.

In 2000, the Five Nations tournament became Six Nations, with the addition of Italy. This means that there is no need for one team to rest each weekend, with three pairs of teams playing each other in a cycle. Sadly this also means that the symmetry of two home/two away matches has been lost; each year half the countries play three home matches, while the other half play only two.

14. FASTEST FINGERS

All three contestants got the same number correct in the same position, meaning two, one or zero letters are in the right place. (It's not possible for exactly three to be in the right place, as that would mean that the fourth was correct too!) All combinations with one or two correct lead to contradictions, so all four got every position wrong. From that we can deduce that the third in the list is not B, D or C so must be A, meaning the second must be C and the correct order is BCAD. Hence a Sentonium has more legs than a Fettlepod.

This puzzle is of course based on the tactics of contestants on the TV show *Who Wants To Be A Millionaire*, which always opened with a 'fastest fingers first' round. With four answers to place in some particular order, the first letter can be any of four, the next is one of the remaining three, and the next one of two, so the number of choices is $4 \times 3 \times 2 \times 1 = 24$. So, if you ever find yourself on the show, and don't trust your general knowledge one iota, a random selection of letters (BDAC is as good as any) has a 1 in 24 chance of being correct. Not great odds, but better than zero.

15. RESHUFFLING THE CABINET

Anerdine: Health to Defence

Brinkman: Chancellor to Education

Crass: Defence to Chancellor

Dyer: Home Secretary to Health

Eejit: Education to Home Secretary

This is what we are told about the moves:

Anerdine → ? → Chancellor

Brinkman → ? → Home Secretary

Crass → ? → Eejit

Dyer → Health

Defence → ? → Education

Since Dyer is moving to Health, neither Anerdine nor Brinkman can be moving to Dyer's job. This means either Crass or Eejit is moving to Dyer's job.

If Crass is moving to Dyer's job, this would mean:

Crass → Dyer (not Chancellor) → Eejit (Health)

Since Anerdine is two steps away from Chancellor, only Brinkman could be before Crass, making the order of moves Anerdine-Brinkman-Crass-Dyer-Eejit-Anerdine, with Dyer currently Home Secretary. However, this would force Anerdine and Brinkman to be Defence and Education, which can't be next to each other.

Hence it must be Eejit moving to Dyer's job:

Eejit → Dyer (not Chancellor) → (Health)

So the order of moves is Crass-Brinkman-Eejit-Dyer-Anerdine-Crass, and in terms of jobs this is Defence-Chancellor-Education-Home-Health-Defence.

This puzzle was published shortly before the British Prime Minister was due to shuffle their Cabinet. Although the story is pure fiction, the reality of what goes on in Cabinet reshuffles is not dissimilar to the situation described in the puzzle. The constraints of ministerial ambitions, egos and personal rivalries mean that a single ministerial sacking or resignation can result in a logical puzzle at least as complex as the one in this story.

In a reshuffle involving five people who switch jobs, the only two outcomes are a 'circle' in which A takes B, B takes C, C takes D, D takes E, and E takes A, or a triangle (A–B–C–A) and a direct swap (D–E). In this puzzle a direct swap was explicitly ruled out, which is why the solution involves a circle: Health–Defence–Chancellor–Education–Home and then back to Health.

Real world reshuffles rarely, if ever, involve circles. Usually what happens is a senior minister resigns or is sacked, and a chain of promotions and sideways moves then happens to fill the void. There are usually several very short chains rather than a single long one. In fact, looking over the last 20 years, despite multiple government reshuffles, we failed to find any chains longer than two (A is sacked, B replaces A, C replaces B). It would seem that the challenge of getting multiple ministers to shift to each other's jobs simultaneously is too much of a headache even for the most politically astute Prime Ministers.

16. AMVERIRIC'S BOAT

If Chestikov does manage to pull the rope horizontally by one metre, then the boat will not only reach the wall but it will start scraping up the side of the quay. This surprises most people!

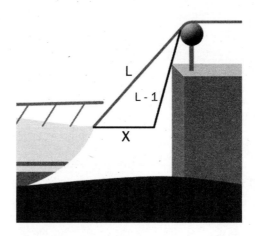

Suppose the boat moves in by a distance X. The taut rope of length L is shortened by one metre to become $L-1$. The two shorter sides of the triangle must combine to be longer than the longest side (that's how triangles work!). So $X + L - 1 > L$, hence $X - 1 > 0$ and $X > 1$.

Most people find this answer surprising. Even maths teachers and academics tend to get it wrong. Why do so many mathematicians get it wrong? Is it something to do with the fact that the horizontal tug of the rope is converted to pulling diagonally, giving the impression that only a portion of the horizontal move is being used? We've certainly found that if you pose this question to people with a strong maths background, they tend to draw on the higher maths that they've learned, and start sketching diagrams with lengths of x and y, and angles marked with Greek letters, and quickly find themselves immersed in messy trigonometrical equations.

Physical experience can improve your intuition. Anyone familiar with mooring a boat will know that pulling the boat in towards a dock is extremely hard work if the rope is at an angle. The reason is that for every metre that you pull the rope, the boat is travelling more than a metre. This is the opposite of pulleys and levers, in which you need less force to lift or pull something because you pull further than the object travels.

17. EXPRESS COFFEE

The location with the smallest combined distance is B. Suppose the depot is lined up with the vendor who is fourth from the left (the 'middle' vendor). If the depot is moved to the left of this, the combined distance of the three vendors to the left will reduce, but the combined distance for the four on the right will increase by the same amount, so the overall combined left-right distance will increase. Similarly if you align the depot with the vendor that is fourth from the top, any change in the vertical co-ordinate will increase the total distance. Kim has realized that the depot should be in the horizontal and vertical position of the middle vendors in those two directions, which is B.

This optimization problem is an adaptation of an old puzzle that's surprisingly little known. In one version dating back 40 years or more, the solution stated that the optimal placement could be achieved by a series of votes: the vendors would be asked if they wanted the depot to be moved North, South, East or West. Voting would continue until there was no majority for movement in any direction, meaning that an optimal solution had been found. This is however a long-winded and perhaps slightly unrealistic strategy.

It's worth pointing out that the optimal depot location is not always a single point. For the location to be a single point there needs to be an odd number of vendors. Think about what would happen if there were six vendors. If you divide them up into groups of three East/West, there is a zone where moving in either direction makes no difference to the total distance (what one side gains, the other loses). Likewise with North/South. In this case, the optimal location of the depot will be a region rather than a point, as the illustration shows.

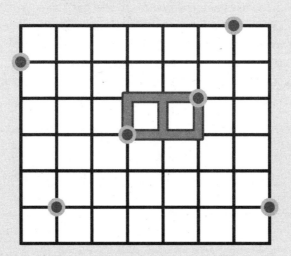

The depot can be placed anywhere on the wider lines shown and the total delivery distance will be the same.

18. SEVENTH TIME LUCKY?

The code is 4321. Septa has four correct digits in six goes. No digit appears in a column more than twice, so she has either got three of the digits correct twice, or two digits correct twice and two once. The digits that appear twice are: 7 (first column); 3 (second column); 2 (third column); 8 (fourth column). The only pair of these that don't coincide are *x* 3 2 *x*. One of the digits 4 8 8 2 is correct, and since 2 can't be repeated it must be 4. Then, from 1191, the last digit must be 1.

Although the puzzle was based on real experience of trying to recall a PIN, some readers expressed amused surprise that Bella was unable to remember a number as simple as 4321, but hey, that's puzzle world for you.

In the real world, personal identification numbers (PINs) are usually assigned randomly by banks, though whether the numbers are genuinely random is unclear. How would you feel if your bank issued you with the PIN 0000? You'd probably think that there had been an accidental reset at the factory. In any case, customers are usually able to change their PIN to a number of their own choosing. And this is where things go wrong, because people are notoriously predictable in the numbers that they choose. Perhaps this isn't too surprising, since we all want a number that is memorable. In 2012, technology consultant Nick Berry analysed over 3 million four-digit PINs that had been exposed online. He found that a staggering 10% of people had chosen 1234 as their PIN. Next most popular were 1111 and 0000, followed by 1212 and 7777 (that last one not too surprising given that the number 7 is comfortably the world's favourite number according to a survey by maths author Alex Bellos in 2014). Our friend 4321 came in at number 18 in the PIN rankings.

What this means is that a credit card thief has a decent chance of hitting on the right PIN just with a bit of trial and error. If repeated digits don't work, try significant dates. In the early 2000s, all PINs in the range 1900 to 1999 featured in the top 20% of the list. No doubt this was because these were birth years of the users (or of their parents or grandparents). These days PINs starting 20xx feature increasingly prominently.

One rather less obvious number that appears surprisingly often is 5683. Why? If your keypad has letters as well as digits, you'll see that 5683 spells out the word LOVE. The least popular number, at least when Nick Berry did his analysis, was 8068, which doesn't have any of the memorable features described above. But presumably by publishing that number it immediately shot up the charts.

19. THE TWO EWES DAY PARADOX

Yes, Farmer Giles is right to be optimistic, though perhaps not for the reason he thinks. The chance that one of the lambs will be female is actually 2/3. We're told that the test detects fragments of the Y-chromosome whenever there is a male. Call the twins A and B. There are three equally likely sex combinations for A and B when at least one is male: M–M, M–F and F–M. In two of these, one of the lambs is female, hence the probability is 2/3.

This puzzle is a variant of a famously controversial puzzle sometimes referred to as the Two Child Problem (and has some similarities with the Monty Hall puzzle, see Puzzle 10).

Here is the scenario. You meet somebody that you don't know, and ask them if they have two children. They say yes. You then ask them if at least one of the two children is a boy. They say yes again. What is the chance that the second child is also a boy?

The classic, surprising answer is that the chance of a second boy is 1/3, not 1/2 as most would expect. Why? Because if I have two children, there is an equal chance that they will be Boy-Boy, Boy-Girl, Girl-Boy or Girl-Girl. There are three equally likely combinations in which at least one child is a boy (BB, BG and GB) and only in the first scenario is the other child also a boy. In 2/3 of the scenarios, the second child is a girl.

There is a related, more mind-bending puzzle called The Tuesday Boy Problem, in which you ask: 'Is at least one of the children a boy born on a Tuesday?', and you receive the answer 'Yes'. This time, astonishingly, the chance that the other child is a boy is 13/27, just under one half. If you want to know where this strange fraction comes from, Google it. We just wanted to highlight the Tuesday in the title of that problem, for reasons that will emerge shortly.

The trouble with Two Child Problems is that the scenario of asking a stranger about the sex of 'at least one of their children' is not how that information would be gleaned in everyday life. For years, puzzlers have been looking for a real-life scenario where one might discover in a natural way that at least one child is a boy without having to ask. Then in 2020, US mathematician Peter Winkler found an example. A friend of his was pregnant with 'fraternal' (i.e. non-identical) twins. As part of the pregnancy monitoring, the mother had undergone an extremely reliable test which always detects Y-chromosomes if at least one child is a boy. The test for Y-chromosomes was positive, so the mother was told that at least one of the twins was a boy. She hoped the other might be a girl, but thought the odds were no better than 50-50.

Because fraternal twins are the same as two independently born children, the sex of each child is a 50-50 random choice, like flipping a coin. And so, in this instance, knowing that one child was a boy, the chance that the other would be a girl was 2/3. As it turns out, the other baby was indeed a girl, so the mum got what she wanted.

We thought this would make a nice puzzle for *New Scientist*, especially with the scientific testing element. However, at a time when binary categorization into boys/girls is a topic of heated public and scientific debate, we decided it would be better if the puzzle could be about animals rather than humans.

After some research, we landed on sheep as the perfect solution. Typically sheep arrive as non-identical twins, and like most mammals, they are roughly 50-50 male/female. We even checked with the world famous Roslin Institute (remember Dolly the Sheep?), to confirm that a Y-chromosome test on sheep could identify that at least one twin was a male, and that intersex sheep are almost unheard of.

We gave the puzzle a story about a rare-breed sheep farmer's prize ewe who is expecting twins. Hoping for another female, the chromosome test reveals that at least one of the twins is male.

And that is how this new version of the puzzle came to be called The Two Ewes Day Paradox.

20. THE HEN PARTY DORM

There was a 50% chance that Janice would end up in her own bed. Anyone who got back and found their bed was occupied picked another bed at random. Two of the unoccupied beds must have been Amy's and Janice's, and there was an equal chance of picking either one. If Amy's bed was taken at any point, then Janice would get her own bed, otherwise Janice would end up missing out. So the answer to what sounds like a difficult probability problem is simply 1/2. The ninth friend, Iona, would get her bed if somebody picked either Amy's or Janice's bed before Iona got back. Since there were more ways in which Iona could end up with her own bed, she was more likely to get her bed than Janice.

This puzzle is usually presented in the context of passengers lining up to board a plane, one of whom has lost their ticket and so picks a seat at random. Traditionally it only asks about the final passenger. We wanted to set the puzzle in a different context. When else would somebody choose the wrong place to go? Alcohol and hen parties, of course.

The question about the ninth friend, Iona, is not usually asked. It's not straightforward to work out the exact chance of her successfully ending up in her own bed, which is why the puzzle merely asks whether she is more or less likely to succeed than Janice. In fact, although Amy (who returned first) only had a 1/10 chance of ending up in her own bed, Beth's (who returned second) chance of success was 9/10 (she would only fail if Amy happened to end up in her own bed). From this, if you guess that each successive party-goer has a lower chance of success than their predecessor, you are right. In fact, the chance of finding the right bed follows a nice pattern: Beth's chance is 9/10, Carol's is 8/9, Dionne 7/8, Eugenie 6/7 and so on, finishing with Janice on 1/2.

21. CUTTING THE BATTENBERG

Cake dissection puzzles have been around for a long time, but so far as we know, the Battenberg variation has added a new twist to the genre. The traditional – some would say slightly sneaky – way to make eight slices is to cut the cake in half, stack one half on top of the other (A), cut through the stack, put one stack on the other (B) and repeat:

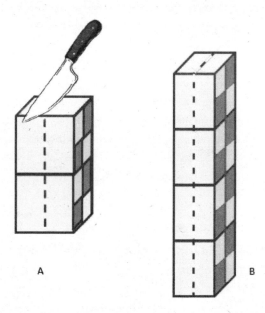

A   B

You now have eight identical square pieces, each with two yellow and two pink squares.

Maths teacher, podcaster and puzzle setter Andrew Jeffrey posed this puzzle on BBC Radio 5 Live, expecting the classic cake-stacking approach to be the unique solution. To his surprise, the production team came back with an alternative that didn't need any stacking.

You can simply cut the cake in half, then set it on its end and cut down both the diagonals to make eight triangular prisms:

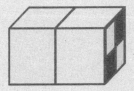

These were the two solutions that were published in *New Scientist*, and as far as we are aware, no others were offered by readers. But a couple of years later, presenting the puzzle to a group of young teenagers, two other solutions emerged – both of them using the cut-and-stack approach.

Cut the cake in half, stack one half on top of the other and cut through the stack to make four pieces (as in A above). Then line up the four pieces as below and either cut horizontally (C) to make rectangular pieces, or along the diagonal (D) to make triangular pieces:

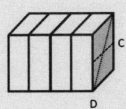 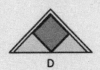

D C D

This leads to more solutions. In the final cut, any line that goes through the central axis of the cake will give you a trapezium-based prism (C and D are just special cases). This means there is in fact an infinite number of solutions to the puzzle!

If you wonder about whether any of these solutions are practical, you can try them out on a real Battenberg cake. The marzipan and jam act as excellent binders, and you should have no problem reproducing the solutions.

22. A JIGSAW PUZZLE

A traditional jigsaw is rectangular, and the number of pieces along the edges multiplied together gives the total number of pieces. The prime factors of 468 are 2, 2, 3, 3 and 13, with 2 and 3 appearing twice. Multiply these five prime numbers together in any order and the result is 468. There are many different ways of combining these factors, for example 2 × 3 (= 6) × 2 × 3 × 13 (=78). We need to find combinations that give plausible dimensions of a jigsaw. Reasonable candidates are: 26 × 18 or 36 × 13 or 39 × 12, etc. But the puzzle states that there was nothing unusual about the jigsaw. Traditional jigsaws are never more than twice as wide as they are high, and 26 × 18 is the only rectangle that is jigsaw shaped. This would have 2 × 26 plus 2 × 16 = 84 edge pieces (including corners).

Some solvers weren't happy with the solution to this puzzle, arguing that the jigsaw could perfectly well have been a 36 × 13, or even a 39 × 12. And in theory they were quite right. However, the puzzle is set in a very informal tone, and the jigsaw in question is an old/traditional one, and featured a picture of a cottage. You would struggle to find an image of such a cottage that is twice as long as it is high. If this were a formal exam question, the setters couldn't get away with allowing for a bit of common sense, but in puzzle world we reckon that (within reason) it's fair.

23. THE NINE MINUTE EGG

Slow method: start egg-timers 7 and 4. When 4 runs out flip it. When 7 runs out there is 1 minute left in the 4. Put egg on; when 4 runs out (after 1 minute) flip it, when it runs out flip it again to get 9 minutes. This means I won't get my egg for 7 + 9 = 16 minutes.

Fast method: start timers 7 and 4 and put the egg on. When 4 runs out (total 4 minutes) flip it; when 7 runs out (total 7 minutes) flip it (i.e. just flip the 7 now, not both); when 4 runs out (total 8 minutes) flip the 7 (which has been running for 1 minute). When the 7 runs out (total 9 minutes) the egg is cooked. So I get my egg in 9 minutes.

There are a variety of puzzles in which you have to measure a precise time using two props. This one used two egg-timers, but there is a better known version involving two timers (lasting 7 and 11 minutes) to measure 15 minutes. (One solution to this one is to turn over the 7 and 11 timers; when the 7 runs out, turn it over; when the 11 glass is empty, the inverted 7 glass has been running for four minutes. Turn it over again, and when it is empty, 11 + 4 = 15 minutes have elapsed.)

Incidentally, short timers (typically less than ten minutes) are usually called egg-timers, and long timers (presumably anything that would result in a rock solid egg) are usually called hourglasses. We are not aware of any formal transition time where egg-timers become hourglasses. Older puzzles along similar lines involve measuring jugs or burning fuses. Perhaps the most famous is the one used in the movie *Die Hard With A Vengeance*, in which two characters, McClane (played by Bruce Willis) and Carver (Samuel L. Jackson) have five minutes to figure out how to measure exactly four gallons of water using a 5-gallon jug and a 3-gallon jug, and a fountain which has an unlimited supply of water. If they fail, a bomb will detonate.

Spoiler alert – here's the solution that the movie's characters came up with. Fill the 3-gallon jug, pour those three gallons into the 5-gallon jug, refill the 3-gallon, and pour two of those gallons to fill the 5-gallon, leaving one gallon in the 3-gallon jug. Empty the 5-gallon, pour the single gallon in there, and then refill the 3-gallon and empty it into the 5-gallon to add to the one gallon that is there, to make four gallons. There's an alternative solution, which we'll leave you to puzzle over.

24. MURPHY'S LAW OF SOCKS

The first sock that I lose is guaranteed to create an odd sock. There are now five socks left: an odd sock plus two pairs. The chance that the next lost sock will create a new odd sock is therefore 4/5, or 80%. Now we have four socks left, two odd socks and one pair, so the chance that the third lost sock will break up the final pair is 2/4 or 50%. The chance of having three odd socks after three washes is therefore $4/5 \times 1/2 = 2/5$ (or 40%). In other words, you will end up with three odd socks 40% of the time, and you will be left with a pair 60% of the time.

Murphy's Law of Socks is a genuine reason why most of us end up with odd socks. However, one reader did question the tactics of continuing to put odd socks in the wash. What's the point of washing a single sock when you've lost the other one? The author defended it on the basis that this is exactly what he does. All his socks are black, but he buys socks with coloured toes. A look through his sock drawer will reveal pairs of black socks with, for example, yellow and blue toes, pink and green toes, and so on.

25. REARRANGING BOOKS

At least seven moves are needed. You can tell this immediately by noticing that seven numbers (1, 3, 4, 5, 6, 8 and 9) are to the right of the next number up (2, 4, 5, 6, 7, 9, 10) and so have to be moved at some point. This rule works for any size of list. There are numerous ways that you can sort the books in seven moves. For example: move 1 between 7 and 2; then 3 between 2 and 6; 4 between 3 and 6; 5 between 4 and 6; 7 between 6 and 9; 8 between 7 and 9; and finally move 10 from the left end to the right.

There is some surprising maths embedded in the apparently simple challenge of re-ordering books. For a start, the method used in the puzzle – of taking out a book and then shoving others to the side – isn't necessarily the most efficient. A better approach can be to simply swap the positions of pairs of books. For example, suppose the books are arranged in the order:

10 2 3 4 5 6 7 8 9 1

The obvious move here is to simply swap books 10 and 1 (a single move). However with book-shoving, the quickest solution is two moves, for example pull out Volume 10, shove the other nine to the left and insert 10 at the end, then pull out 1 and shove the left-hand eight to the right. That requires two moves, and a lot of strenuous book-shoving. But sometimes shoving requires fewer moves than swapping. If the books are ordered 10 1 2 3 4 5 6 7 8 9, only one shove is needed to sort the order, whereas nine swaps are needed with book swapping, assuming your strategy is to get book 1 into the correct position, then 2, 3, etc.

And sometimes the two approaches need the same number of moves. The books in the puzzle were ordered 10 7 2 6 5 4 1 9 3 8, which can be done in seven swaps or seven shoves.

There's another approach that isn't necessarily efficient, but which you could use if you were (say) particularly weak and short-sighted and could only compare and swap neighbouring books. With this technique, you start from the left with the first and second books and swap them if the left number is larger than the right. Then do the same with the second and third books, third and fourth, and so on, repeating as necessary. With the book arrangement that was used in the puzzle, it turns out that this would take 29 steps. In the worst-case book arrangement of 10 9 8 7 6 5 4 3 2 1 it would take $9+8+7+6+....=45$ swaps, compared with just five swaps if you use the earlier swapping method. But for 10 1 2 3 4 5 6 7 8 9 the two swapping techniques are equally efficient.

The swapping-neighbouring-pairs method is one of the simpler sorting algorithms that computers use. It's known as a 'bubble sort'. Needless to say, computer scientists have come up with some ingenious algorithms that are far quicker than bubble sorting, but for small data sorting jobs it's fine. However, what works with computers doesn't necessarily work with books. Humans sorting books find it easy to pick out the high- and low- numbered books – it's physically swapping the books that takes the time and effort. In contrast, the task of swapping bits of data is trivial for a computer; it's comparing pairs of values that takes all the time.

26. HIDDEN FACES

The opposing faces of a dice always add up to seven. Therefore the face presented by the dice that is on the extreme right must be 5.

The central five dice contribute another 5 × 7 = 35 to the total, regardless of their orientation.

The dice on the furthest left has 6 on top and 1 on the bottom, and by comparing the orientation of the 2 with that on the dice at the right, the furthest-left dice must present a 4 to the adjoining cube. (You might want to get real dice to help you visualize this!)

Hence the total of the touching faces is 5 + 35 + 4 = 44.

The opposite faces of a dice don't have to add to seven. It would be just as 'ordered' if the opposite sides were 1 and 2, 3 and 4, and 5 and 6. Yet you'd struggle to find a cubic dice anywhere in the world that doesn't follow the sevens rule. The sevens rule actually dates back to ancient Egypt and was followed almost without exception by the Greeks and Romans, and nobody is quite sure why. One theory is that it appeals to a sense of symmetry and balance to have each pair of opposites adding to the same total, though since the numbers were always carved out of the face rather than printed, the dice isn't exactly balanced: the 'one' side is heavier than the 'six' which will ever so slightly increase the chance of the dice landing with 'six' on top. Another theory behind dice numbering is that the sevens pattern tied into the historic mystical importance of the number seven, which is still to be found in the number of days in the week, notes in a music scale, colours in rainbow and elsewhere.

There is a neat little magic trick that relies on the seven property. Cover your eyes and ask your volunteer to roll three dice and to add up the three numbers on the top. Now ask the volunteer to pick up any of the three dice, look at the number on the bottom of it, and add that to the running total. Now ask them to re-roll the dice they are holding, and add the new number on top of that dice to the running total. Open your eyes, look at the dice (let's say the numbers are 1, 4 and 5) and explain that you have no idea which dice was picked up, yet you know that the total that your volunteer is thinking of is... 17.

How does it work? The secret number is the total of the three dice you are looking at (10 in this case) plus the numbers on the top and bottom of the dice that was re-rolled (which will always be 7).

27. BIRTHDAY CANDLES

If you start by blowing on one candle and work your way around the cake, blowing on each candle once, this will put out all the candles in seven puffs - and this is the fewest puffs you can do it in. The trick is to make sure each candle gets blown on an odd number of times, which is what this method achieves.

Katie Steckles was looking for a puzzle that would fit with Hannukah and realized there was interesting maths in the lighting and blowing out of candles. As far as we know, trick candles that toggle on and off with each blow don't really exist, though it feels like they should.

For circular cakes, the minimum number of blows required to extinguish N trick candles is always N (most easily done by going in a cycle around the cake), unless N is a multiple of 3 (call it 3k). If N is 3k, only k blows are needed because you simply blow out three candles each time.

For Hannukah, the candles are actually arranged in a straight line rather than a circle. They are set in what's called a menorah, which has four candles on each side of a raised 'shammash' candle in the centre. A straight line of candles makes the blowing out puzzle much simpler, in that you can always blow out all N candles in N/3 blows, with N/3 rounded up to the nearest whole number. With nine candles, the menorah would only need three blows to extinguish them all.

Different versions of the puzzle can be found in other settings. For example, at the mathematical visitor centre MathsCity in Leeds it has been set up as a series of buttons set into a table that control LEDs around the outside.

28. LIGHTBULB MOMENT

Only two lightbulb flashes and one climb of the stairs are needed.

Join two wires at the bottom (e.g. A and B). Climb the stairs. Identify which pair at the other end make a circuit (e.g. 1 and 3).

Now connect either of the pair of top wires that make a circuit (e.g. 3) to a new wire (e.g. 4) and go back down to the ground floor. We already know that the other end of 3 must be one of the wires that made the circuit (in this case A or B), and the other end of 4 will be C or D. Find which pair now makes a circuit (e.g. A and D). You now have all the information you need to identify all four wires.

For example, if the second circuit is A and C with 3 and 4, then A must be 3, so C is 4, B is 1 and hence D is 2.

Light switch/bulb puzzles have long been popular with puzzlists. The classic one involves three switches in the basement and three bulbs on the top floor. You are challenged to identify which switch controls which lightbulb, but this time you have to do it without descending the staircase. It seems there isn't enough information: a light switch only has two states, on or off, and this is insufficient to discriminate between three lights. The 'aha' lateral thinking step is to realize that as well as being on or off, a lightbulb can also be hot or cold. Turn on switch A and switch B, leaving switch C off. After five minutes, turn off switch B. When you get to the top floor, touch the two bulbs that are not lit. The warm one is operated by switch B, and the cold one by switch C.

29. BLOXO CUBES

The six yellows and three blues will fit. The blues are placed along the diagonal of the 3 by 3 box. The diagram shows the box part filled. The three remaining yellow blocks will slot in to the left, right and top faces.

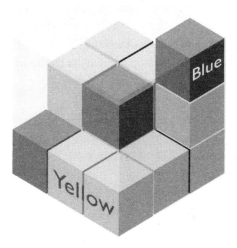

The setter Katie Steckles first encountered a version of this puzzle on the wall of a classroom in Indonesia, though it probably originated in the USA. Realizing that you can't have more than three of the 2 by 2 pieces lying horizontally, and that if more than two of them are horizontal you can't fit the others in, quickly reduces the number of options that you need to explore. You can get quite far just imagining abstract blocks (but it's obviously much more fun to solve if you can get your hands on some).

30. KNIGHT NUMBERS

Knight 618, Bishop 591.

Call the knight numbers BCD and ABCD, and the bishop numbers XYZ and AXYZ. Number AB must be different from AX because the knight and bishop cannot type the same consecutive digits. This means YZ must be high enough that adding 27 increases the hundreds digit. Since there is no zero on this number pad, the only option is that Y = 8 or 9 and C = 1. The only four-digit numbers a knight can type with 1 in the tens place are 3816 and 7618. Subtracting 27 gives us 3789 and 7591, and of these only 7591 is a bishop number. So the numbers that the son found are 591 and 618, and I could have added a 3 or 7 at the front.

The son of Peter Rowlett, the setter of this puzzle, really did develop an obsession with chess. He was typing the answer to a multiplication question, 3×9, into the computer and shouted excitedly that the answer was a knight's move on a telephone keyboard. That made Peter think of the knight's tour puzzle. This asks for a sequence of moves so that a knight visits every square on a chess board exactly once and, in some versions, returns to the original square. A knight's tour is not possible on a 3 by 3 numberpad because there is no way for the knight to reach the central number 5 from the other numbers. But if you remove the 5 key, it is possible for a knight to jump around the other keys.

The prolific Victorian puzzlemaker Henry Ernest Dudeney solved problems of this sort using what he called a 'buttons and string' method. Here, he would represent each space on the keypad (except 5) using a button. Then he would connect two buttons with a piece of string if a knight could jump from one to the other in a single move. The result appears a rather tangled mess, but if the buttons are moved around then the picture becomes a lot clearer, as shown in the diagram that follows. Now you simply choose any key as the starting position and move the knight round the circle in one direction or the other to find two different knight's tours.

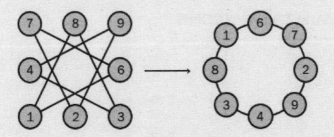

Nowadays we recognize this as a topological argument, part of a mathematical topic called graph theory – with the buttons called vertices and the string as edges. A route around a graph that visits every vertex exactly once is called a Hamiltonian path, whereas a route that passes over every edge exactly once is called an Eulerian path (you'll find more about these in puzzle 55, Dungeons and Diagrams). As it happens, both concepts have their origins in mathematical puzzles, and these ideas have applications in route finding, computer circuit design and DNA sequencing.

Chapter 4 **FIGURING IT OUT**

31. SUM THING WRONG

The real sum is:

$$
\begin{array}{r}
1\ 6\ 5\ 6 \\
+\ 1\ 0\ 7\ 6 \\
\hline
2\ 7\ 3\ 2
\end{array}
$$

It helps to replace the original wrong digits with letters, so:

$$
\begin{array}{r}
A\ B\ C\ B \\
+\ A\ D\ E\ B \\
\hline
F\ E\ G\ F
\end{array}
$$

Since 2B ends in F, F is even, so in the left-hand column 2A = F.

Since every digit in the new sum is different from its partner in the original sum, we know that A is not 2, so it can only be 1, 3 or 4, meaning F is 2, 6 or 8, and B (which ends in F when doubled) is 6, 8 or 9.

But B + D must be less than 10 to avoid carrying 1 to the left column, so B cannot be 8 or 9, meaning B = 6, A = 1, F = 2 and the other numbers follow.

This 'wrong sum' puzzle is a twist on a classic form of puzzle known as a cryptarithm or alphametic. In these arithmetical puzzles, the digits in a calculation are replaced by letters. A basic cryptarithm is one of the easiest puzzles to set. Take any calculation – for example 23 + 87 = 110 – and substitute each digit with its own unique letter, in this case (say) AB + CD = EEF. That's it, the puzzle has been set. The challenge for the solver is to work out what the original sum was. This particular example isn't very good though. There are 128 legitimate solutions to AB + CD = EEF (for example, 39 + 76 = 115 would also fit the pattern). To be satisfying, a cryptarithm should have a unique solution, and a method of solution that requires some insights rather than pure trial and error. And the very best cryptarithms aren't just a random assortment of letters, but are made up of words that make sense. The most famous was posed by H. E. Dudeney, which might have been a telegram sent by a student to their parent: SEND+MORE=MONEY. It has a unique solution (see footnote).

In the Sum Thing Wrong puzzle, the original digits have been replaced by digits rather than letters. This constrains the number of solutions, because each digit in the puzzle removes itself as a possible solution (we know that '6' can't be a '6'!). A perfect digits-for-digits puzzle would be a sum that is correct in both its original and its 'encrypted' form. That's not the case in the puzzle set here, though the fact that the sum is wrong has been built into the story. As a bonus, although the sum is wrong, it is plausibly wrong: the first and last columns suggest that the sum might be correct, until you do a bit more investigation.

9567 + 1085 = 10652

32. CAR CRASH MATHS

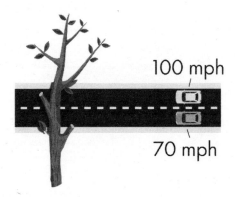

The surprising answer is that the yellow car hits the tree at 70 mph - the same speed that the blue car was going when the driver first spotted the tree. The energy expended on braking is the force multiplied by the distance. If we assume that the force applied on the brake was the same for both speeds (foot down on the floor!), then since the distance travelled was also the same, both cars expended the same amount of energy. The energy of a moving object is given by the formula $1/2 \times m \times V^2$, where m is the mass and V is the velocity. The blue car braked to a halt, so its starting velocity was 70 mph and its ending velocity was 0 mph. This means it lost $1/2 \times m \times 70^2$ of energy. The yellow car started with $1/2 \times m \times 100^2$, and by braking it lost the same amount as the blue car: $1/2 \times m \times 70^2$. Since 70^2 is roughly half of 100^2, the yellow car still has the other half of its energy, i.e. it is now travelling at 70 mph.

This puzzle is an exception to the general rule in the *New Scientist* column in that it requires some school mechanics knowledge to solve it - but we reckon it was justified by the surprising answer.

Several years ago, when the setter of this puzzle, Ben Sparks, was still a maths teacher, there was a TV campaign called 'It's 20 for a reason', about speed limits near schools. The advertisement compared the stopping times at 20 mph and 30 mph. One teacher in the staffroom wondered what the equivalent was at motorway speeds. The sobering punchline made it a compelling puzzle with a thought-provoking result.

How can drivers develop better intuition about the significance of driving at different speeds? One of Ben's suggestions is to have a non-linear speedometer, in which the gaps between speeds are proportional to the square of the speed. That way, driving at 100 mph would look like it is double the speed at 70 mph – which in stopping distance terms, it is.

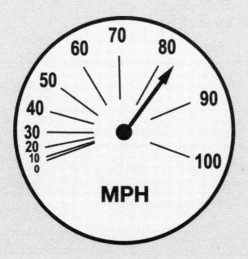

A speedometer that gives a better sense of danger.

33. SOCCERDOKU

Every team played three matches, hence United drew one, and Albion won one but drew none. United conceded 0 goals, so they drew with Town 0-0, and hence beat Rovers 1-0 and Albion 1-0. The other results follow to give:

United	1	Rovers	0
United	1	Albion	0
United	0	Town	0
Rovers	2	Town	0
Rovers	2	Albion	0
Albion	3	Town	0

Do-it-yourself 'Soccerdoku' puzzles can be found during any international soccer tournament, such as the World Cup or the Euros, in which the initial stage involves mini leagues of four countries. It is often possible to deduce the match results even when all the round robin matches have been played. For example, this was how Group C ended in the 2018 FIFA World Cup.

	P	W	D	L	F	A	Pts
France	3	2	1	0	3	1	7
Denmark	3	1	2	0	2	1	4
Peru	3	1	0	2	2	2	3
Australia	3	0	1	2	2	5	1

If you can't remember the results, a little thought, deduction and trial and error will enable you to work them out from the table. France and Denmark were unbeaten so Peru must have beaten Australia. Since Peru lost twice and only conceded two goals, they must have lost 1-0 in two matches (against France and Denmark) and beaten Australia 2-0. Denmark's results were therefore 0-0 and 1-1, and... well, you can figure out the rest (or check the original table). By the way, France went on to win the World Cup that year.

34. ALL SQUARES

a. Natalie was born in 1980. The only year in the current century that is a square number is 2025, which is 45^2. (The next square year after that will be 2116.)

b. The difference between two squares, $a^2 - b^2$, is (as any older school maths student should know) the same as $(a + b)(a - b)$. This means the original question can be written as $[(68 + 32)(68 - 32)]/[(59 + 41)(59 - 41)]$.

$68 + 32 = 59 + 41 = 100$. Therefore $(68^2 - 32^2)/(59^2 - 41^2) = (100 \times 36)/(100 \times 18) = 36/18 = 2$.

The square age curiosity has been exploited in niche quiz questions such as: 'Which two Wimbledon tennis winners will celebrate their Nth birthday in the year N^2?' This means looking for champion tennis players who were born in 1980 and will be 45 in the year 2025. You might think any such winners must have been in their 20s or maybe 30s when they won, and in the case of Venus Williams you'd be right. But there was also a teenage winner born in 1980: Martina Hingis won Wimbledon in 1997, a little before her 17th birthday.

The formula for the difference of two square numbers lends itself to a nifty mental arithmetic trick. Announce that you know how to square any number between 10 and 99. Let's say your friend gives you the number 97. To square the number, find the nearest 'easy' number to it – in this case 100 – and note the number B that you needed to add (in this case $B = 3$). Subtract B and multiply to get 100 $\times$ 94 (=9400), then add the square of B (=9), and the answer is 9409. When you are confident in the technique, you can try it on harder numbers: $88^2 = 100 \times 76$ (=7600) and add 12^2 (=144) to get 7744. With a bit of practice you've now got a great party trick. As long as you choose your parties carefully.

35. SQUAREBOT

Since Squarebot was kind enough to draw on squared paper, we can assume that the dimensions are integers. So we might ask 'What's the square of the height?' and Squarebot will give an exact answer. Unfortunately, if Squarebot says 289, the shape could be a 17 × 17 square or a 17 × 18 rectangle. But what will always work is asking a question where the answer for a square would be zero, such as 'What's the difference between its height and its width?' If it's not a square, Squarebot would give an answer of at least 1.

Catriona Agg, who set this puzzle, is best known for her hand-drawn geometry puzzles that have achieved something of a cult status on Twitter. Her normal process of inventing puzzles is through the process of scribbling in her notebook, but the Squarebot puzzle occurred to her while out for a walk without any notebook at hand. She was thinking that her favourite geometric puzzles are ones where the answer can be found from surprisingly little information, for example, when you can calculate an area of a shape without knowing all of the dimensions involved. It occurred to her that a similar principle is often used in logic puzzles. She describes the Squarebot puzzle as what you find at the intersection of a geometry and a logic puzzle.

36. CHRISTMAS GIFTS

a. I have received 42 geese-a-laying and the same number of swans-a-swimming. The pattern goes 12 × 1 (partridges), 11 × 2 (turtle doves), 10 × 3 (French hens) and so on, reaching a maximum at 6 × 7 and 7 × 6.

b. The total number of gifts I received was 1 + (1 + 2) + (1 + 2 + 3) + ... + (1 + 2 + 3 + 4 + 5 + 6 + 7 + 8 + 9 + 10 + 11 + 12). Add that lot up and you get 364. If I give away the first gift on the 26th December, I give away the last gift 363 days later. In most years that would be Christmas Eve, but in 2019 when this puzzle was set, the following year was a leap year, so the gifts lasted until 23rd December.

It is easy enough to calculate how many gifts 'my true love' receives each day. The numbers 1, 3, 6, 10, 15, etc are known as triangle numbers, which you can visualize in the form of balls arranged into triangles. (A regular game of pool begins with 15 balls in a triangle.)

To find the Nth triangle number use the formula: $1/2 \times N \times (N + 1)$, so on the 12th day, my true love gave me $1/2 \times 12 \times 13 = 78$ gifts.

What is less well known is the formula for the sum of the triangle numbers. For the record, it is $1/6 \times N \times (N + 1) \times (N + 2)$. These are known as the tetrahedral numbers, which you can picture as triangles of balls stacked on top of each other, as used to happen with cannon balls back in the days of the Napoleonic Wars:

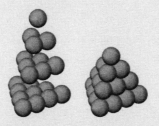

For $N = 12$ the formula produces the answer $1/6 \times 12 \times 13 \times 14 = 364$. It is a delightful coincidence that this happens to be one less than the number of days in a year, which means I have one present to enjoy every day between Boxing Day and Christmas Eve, except during leap years.

37. TIGHTWAD'S SAFE

The code of the safe is 9075461328. The most useful clue is that the digits 0 to 9 are all used along the top and side of the grid. 0 times anything is 0, so the second column must be 0. One times anything can only have one digit, so the only possible place for 1 is the second row. Two times anything is even and is less than 20, so the only place left for 2 is the fourth row. And since three times something is less than 30 the third row is the only place left for 3. Then, 5 must be the fourth column since one of the numbers ends in 5, and similar logic leads to all of the other digits:

	9	0	7	5	4
6	54	0	42	30	24
1	9	0	7	5	4
3	27	0	21	15	12
2	18	0	14	10	8
8	72	0	56	40	32

This puzzle proved particularly popular with maths teachers, who picked it up on Twitter. It's a satisfying exercise in factorising numbers, and turned out to be directly usable in the classroom. Some schools used it as their Puzzle Of The Week.

38. THE CARD CONUNDRUM

Bob's solution was $x = 21$. The formula was written on a card and when Alice picked up the card she must have looked at it upside down. The upside down equation $8/(x-1) - 1 = 9/x$ has two solutions: $x = 3$ and $x = -3$. These can be found with some trial and error, or (if you recall your school maths) by rearranging the equation and cancelling, which results in $x^2 - 9 = 0$, to which the solutions are $x = +3$ and -3.

The games and puzzles YouTuber Tim Rowett, famous for his 'Grand Illusions' channel, had this equation written out on the back of a business card that he showed to one of us at a gathering. In that form, it was a curiosity rather than a puzzle, but we tweaked it to make the upside down property an 'aha' discovery. It relies of course on the fact that the numbers 1, 6, 8 and 9 are all still digits when turned upside down, as is zero. Mathematician Peter Rowlett (resemblance to the name Rowett is a coincidence) looked for other equations that are still solvable upside down. He hoped to find equations that were different when inverted but had the same solution either way up. The best he found was $(x+1)/6 = 9/(8+x)$, which has two solutions, roughly 3.64 and −12.64 (alas, neither of them whole numbers). A handwritten 5 can be made to look like a 5 when upside down, which opens up other equations that work, such as $(x+5)/9 = 6/(8+x)$ with the solutions 1 and −14, both ways up.

39. MARTIAN FOOD

Let's call the sustainable population of Mars P.

P people consume the amount of grass that grows in any period.

40 - P people consume the remaining stock in 20 days.

20 - P people consume the remaining stock in 60 days.

So: $20 \times (40 - P) = 60 \times (20 - P)$

And hence $P = 10$.

This is a modern and greatly simplified variant of a puzzle first set by Isaac Newton over 300 years ago. If you don't think of Newton as being a fun puzzles kind of guy, you'd be right. The puzzle was in the form of a mathematical problem in the book *Arithmetica Universalis*, which was an edited version of Newton's lecture notes. The problem was not accessible to the general public because the entire book was written in Latin. The other deterrent in Newton's version was that it contained some rather awkward numbers. The original puzzle (translated into English) went as follows:

In 4 weeks, 12 oxen graze bare 3 1/3 acres of pasture land, and in 9 weeks, 21 oxen graze bare 10 acres of pasture land. Accounting for the uniform growth rate of grass and assuming equal quantities of grass per acre when the pastures are put into use, how many oxen will it take to graze bare 24 acres of pasture land in a period of 18 weeks?

How do you solve this? One way is to set up names for different variables, for example: *A* for the number of acres, *O* for oxen, *W* for weeks, *G* for the rate at which oxen eat grass and *R* for the rate at which grass grows. Next set up a series of equations to reflect the grazing in the three different situations in the problem and then... you know what, we suspect most readers of this book don't want to get bogged down in solving three simultaneous equations. However, if you can get your head around that lot, you should eventually emerge with the answer 36 oxen – and Newton would be proud of you.

40. DIFFY

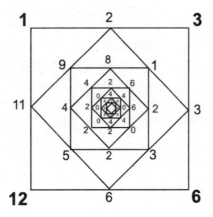

The numbers 1, 3, 6, 12 clockwise around the square will give you ten Diffy squares. It is possible (though not easy) to find the four numbers by a combination of working backwards and thoughtful experiment. For example, since the final square has O O O O at its corners, the square before it must be A A A A (where A is some whole number), and before that must come O A O A and before that, O A A O. Beyond this some more experiment and trial and error is needed.

If all the starting numbers on a Diffy square are integers then you will always end up with four zeroes. Why? Because after each step, the total of the four numbers must get smaller. The minor exception is if you start with negative numbers, e.g. –1, –5, 10, 20. Since 'difference' means 'the higher number minus the lower', after the first round you end up with four positive numbers whose total is higher (4, 15, 10 and 21 in this case). However, after that initial increase, the total diminishes after every subsequent step.

The maths behind Diffy gets quite deep very quickly, but it turns out that any combination of numbers, including fractions and irrational numbers such as π, will end at four zeroes, often after only a few iterations. There is, however, an exception: one particular combination of numbers that never reduces to zero, but instead goes on forever, leading to what has been nicknamed the Diffy Zombie. The four numbers at the corners of the Diffy Zombie are (approximately) 1, 1.84, 3.38 and 6.22, though the actual numbers are irrational numbers. (If you're wondering where they come from, they are derived from solving the cubic equation $a^3 - a^2 - a - 1 = 0$... but if you want to delve into this subject, look up 'Ducci sequences' online.)

Numbers that create a high Diffy score are related to what is known as the 'Tribonacci sequence': 1, 1, 2, 4, 7, 13, 24, ... , in which each term is the sum of the previous three. Tribonacci is a pun on the more famous 'Fibonacci' sequence in which each term is found by adding the previous two numbers. Note that the four Tribonacci numbers 2, 4, 7, 13 are one higher than our solution 1, 3, 6, 12.

41. PIECES OF EIGHT

The time in bed is 8 hours 57 minutes.

Each of the four digits on the alarm clock is made up from a conventional seven-segment display.

The 'O' uses six segments, the '1' uses two, etc. so the digits, from dimmest to brightest are:

1, 7, 4, (2, 3, 5), (0, 6, 9), 8

with equal brightness in brackets.

The four clock digits can be 0-1, 0-9, 0-5, 0-9.

So the dimmest display is 11:11 and the brightest is 08:08.

The idea of creating numbers out of a seven-segment display dates back to the year 1903, and possibly earlier. However, these numbers didn't gain mass appeal until the arrival of digital watches with LED displays in the 1970s, which is when they started to feature in puzzles. Displays now have a standardized form like this:

However, it took a while to agree whether 6 and 9 should have a horizontal tail at the top/bottom (these days they always do), whether 7 should have a vertical segment top left (these days it doesn't), and whether the 1 should use the segments on the right or the left (right became the standard).

42. WHICH FLIPPING YEAR?

The date with the biggest flipping difference was 1066, the year of the Norman conquest of Britain by William the Conqueror.

The calculator digits that are 'flippable' are 0, 1, 2, 5, 6, 8 and 9, though 6 and 9 turn into each other on rotation. 9901 - 1066 = 8835.

Those of a certain age will remember having fun rotating calculator displays to reveal words rather than numbers. 710.77345 to create ShELL OIL used to be a popular one. There were a couple of rude ones as well. Alas, modern calculators have much more sophisticated displays, with numbers that are curved to resemble those found in regular fonts. The digits 2, 4 and 5 no longer resemble numbers or letters when rotated, depriving school children of a source of distraction.

43. SIX WEEKS OF SECONDS

The numbers are the same! The number of seconds in six weeks is 6 weeks × 7 days × 24 hours × 60 minutes × 60 seconds = 6 × 7 × (3 × 8) × (5 × 4 × 3) × (5 × 2 × 3 × 2 × 1), which can be rearranged to 10 × 9 × 8 × 7 × 6 × 5 × 4 × 3 × 2 × 1.

If, like many people, your school summer holiday lasted six weeks, then you already have a good sense of what that length of time feels like. This, when put alongside the coincidence that 10! is equal to exactly 6 weeks of seconds, makes the puzzle a very pleasing, easily checkable, and genuinely curious fact. It is nothing more than a nice coincidence, which arises from our arbitrary definitions of the units of time.

Another curiosity that arises from basing our time system on 60 and 24 is that the number of milliseconds in one day is equal to: $5^5 \times 4^4 \times 3^3 \times 2^2 \times 1^1$ (= 86,400,000 since you ask).

Finally, if you've ever wondered how many minutes there are in February, the answer is 8! That's 8 x 7 x 6 x 5 x 4 x 3 x 2 x 1, which comes to 40,320.

44. LARA'S BIRTHDAY

Francesca and Martha's birthdays are 29th Feb and 1st March. Those born on the 1st were 1 on their first birthday, but not 31 (etc) the day before. Meanwhile, nobody can celebrate their 29th birthday on the 29th February because their age on that date can only be a multiple of 4.

This is an example of a puzzle that emerged directly from a real-life incident. The setter found out that a teacher colleague was turning 29 on the 29th of the month, and initially just thought it was a pleasing coincidence. But he naturally wondered whether this must happen to everyone at some point in their lives, provided they lived at least as many years as the number of their birthday date. A discussion in the staffroom ensued, which set him off on a new train of thought about leap years. For example, it seems odd on first hearing it that nobody ever has their 21st birthday on the 29th February, but a moment's thought will reveal why that is the case (Feb 29th babies only have birthdays on years that are a multiple of four).

It reminds us of another birthday oddity. One day Erica announced: 'a couple of days ago I was 35, next year I'll be 38.' When was her birthday?

Answer: Erica was born on 31st December and today is 1st January. Two days ago, on the 30th, she was 35; yesterday, her birthday, she was 36; on her birthday this year she'll turn 37, and on her birthday next year, she'll turn 38.

45 THE MOUNTAIN PASS

The hikers both set off at 6 a.m. Suppose they cross after X hours. Bonnie has four more hours to go and Aaron has nine. The distance Aaron has travelled is equal to the distance that Bonnie is yet to travel. Assuming (as we must) that they both have constant walking speeds then it helps to think in ratios. Aaron and Bonnie will each cover the same distance in the afternoon as the other one covered in the morning.

After X hours:

Bonnie morning   Aaron morning

The time Bonnie has travelled divided by the time she has left to travel is $X/4$ and for Aaron the time he has yet to travel divided by the time he has travelled is $9/X$. These two are equal, hence $X^2 = 36$ so $X = 6$

In this puzzle, the two characters walk at constant speeds but reach their destination at different times. There is another mountain path puzzle which is the other way round – the same start and end times but different walking speeds. Usually the problem is presented as a story about a monk.

On Monday, the monk leaves the monastery at the foot of a mountain at dawn, climbing steadily with barely a pause for breath. At dusk he reaches the summit, where he spends the night in contemplation. On Tuesday he sets off from the summit at dawn at a brisk pace. He stops to admire the views a few times, and around lunchtime he takes a nap. He arrives back at the monastery at dusk. The question is: can you be certain that there was a time on Monday and Tuesday when the monk was at exactly the same point on the path at the same time of day?

There is an elegant solution to this. Imagine filming the monk on both days, using a fixed camera that can see his entire route. Now superimpose the two films. At dawn, the Monday monk is at the foot of the mountain and the Tuesday monk is at the summit, and at dusk their positions have reversed. There must be a point where the two monks pass each other on the path, and that is the moment when they were at the same place at the same time.

46. A WELL-TIMED NAP

One hour is 60 minutes on the minute hand, and one twelfth of a circle = 30° for the hour hand. In other words, if the time is m minutes past the hour, then the hour hand has travelled through $1/2 \times m$ degrees past the hour.

Suppose that when I go to sleep, the angle between the minute and hour hand is m. If the minute hand is just 'ahead' of the hour hand this means $(30 - 1/2 \times m) = m$. This is only true for $m = 20$, meaning the time is 3:20 p.m. It's also possible that the minute hand is more than one hour segment ahead of the hour, i.e. $30 + (30 - 1/2 \times m) = m$, meaning $m = 40$, and the time is 6:40 p.m.

There are no other solutions. $60 + (30 - 1/2 \times m) = m$ means $m = 60$, and 60-past is not a recognized time, nor is 12 o'clock in the afternoon. If the hour hand is *ahead* of the minute then $(30 + 1/2 \times m) = m$ and again $m = 60$, which doesn't give a time in the afternoon.

So I fell asleep at 3:20 p.m. and woke at 6:40 p.m., having slept for 3 hours and 20 minutes.

The familiarity of a clock face makes it a good basis for a puzzle. There are a number of simpler puzzles that can catch people out. For example:

(a) between one minute past noon and one minute before midnight, how many times does the minute hand exactly overlap the hour hand? And related to that...

(b) in a single day, how many times are the minute and hour hand at right angles to each other? And perhaps simplest of all...

(c) if you have a clock which has ticks instead of numbers to indicate the hours, and you turn that clock upside down, would you always be able to tell that something was wrong?

Spoiler – here are the answers.

(a) The hands overlap ten times (not eleven). It first happens at just after 1:05 p.m., then a little after 2:10 p.m. and so on up to a bit after 10:54 p.m. After 11 p.m. the minute hand doesn't catch up with the hour hand until midnight.

(b) The hands are perpendicular 44 times in a day. The two obvious times are 3 o'clock and 9 o'clock (a.m. and p.m.), but there are other fiddly times when the hands are also perpendicular. It first happens at around 12:16 a.m., then at about 12:49 a.m., 1:21 a.m. and so on. It appears to be twice per hour, but after 7:54 (ish) and 1:54 (ish) the next times are 9:00 and 3:00.

(c) An upside down clock will never tell the right time. The minutes have advanced by half an hour, but the hour hand is still in the same position between the hours, so it is now half an hour early. For example, if you turn a clock upside down at 3 o'clock, the minute hand is at half past, but the hour hand is pointing at the 9, whereas it should be midway between 9 and 10.

47. TRIPLET JUMP

Let's call the mother's age when the triplets were born M. After X years, the mother is $M + X$ years old and the triplets are X years old, and we're looking for a time when $M + X = 3X$, in other words when $M = 2X$. Since X is a whole number, M must therefore be even. So the triplets can only have the same birthday as the mother if her age when they were born was an even number. Since her age when giving birth was equally likely to be even or odd*, the chance that the triplet coincidence will happen is 50%.

It's extremely unlikely that a mother and her triplets would share a birthday. What if (as is far more likely) the triplets had not been born on their mother's birthday? The mother was M when they were born, and had her $(M + 1)$th birthday before the triplets' first birthday. When the triplets reach X years old, there will be a period when the mother's age is $M + X$, and for the rest of the year she will be $M + X + 1$, so the coincidence could happen when either $M + X = 3X$ (i.e. $M = 2X$) or when $M + X + 1 = 3X$ (i.e. $M + 1 = 2X$). So regardless of whether M is even or odd, there will always be a value of X that works. The coincidence is guaranteed to happen at some point!

This puzzle was inspired by a genuine request from a friend, who emailed in to report that her three children had observed that their ages all added up to hers. 'What are the chances?' she'd asked. Since they all had different birthdays, it turns out that it was a certainty that this would happen at some point in her life.

---

* This isn't precisely true – women are more likely to give birth at some ages than at others, but regardless of the country or culture, the split between even and odd ages is likely to be very close to 50-50.

48. HALF TIME

The numbers on the two pieces added to 69. The numbers 1 to 12 add to 78, but Holmes had observed that one piece added to an odd number and the other to an even number: odd plus even equals odd, and cannot equal 78. The only way to achieve two pieces that together add up to an odd number is if one of the two-digit numbers breaks so that its two digits end up on different pieces. This could be 10 (splitting to make 1 + 0), 11 (1 + 1) or 12 (1 + 2). In each case the total of the numbers has decreased by 9, and 78 - 9 = 69.

This is another puzzle that owes its origins to H.E. Dudeney, the recreational mathematician and puzzle author. In the original, the task was to break a Roman numeral clock face into four distinct pieces such that the sum of the numbers on each piece added up to the same number. Here's a clock face, for reference.

It turns out that there are multiple paths to the solution, but, as mentioned in the solution above, the numbers on a clock add up to 78, which is not evenly divisible by four. So the answer requires you to split the numbers into pieces. Below is Dudeney's solution:

In an attempt to find a similar puzzle that would work for the more modern clock face with Arabic numerals, Brian Hobbs was at first disappointed to realize that splitting the 10, 11 and 12 all had the same effect on the overall sum, but then realized that, when framed differently, it made for a nicely succinct puzzle with an 'aha' moment that converges on a single solution. Or, almost. Somebody sent in this unlikely, but certainly possible alternate solution to Half Time, which would make the resulting sum 51.

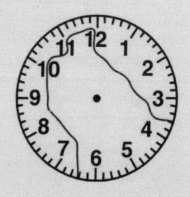

49. SEEING RED

If I see a green light, I should ease off as I will miss the light. If I see a red light, I should accelerate. The actual speed and distance don't affect the answer, but suppose I'm riding at 4 m/s. The lights are then 60 metres away when I first see them.

If they're green then I can speed up by 25% to 5 m/s but there's really no point - I'll certainly miss them by at least 10 metres. That's life, no point in getting cross.

If they're red then it's best to go at 5 m/s for 4 seconds. That way I'll be 40 metres away, close enough for when the lights change to green. If the lights change during those 4 seconds when I've bothered to speed up then I'll be 'really cross' because I'm still too far away. The total lights sequence is 100 seconds so on average that's going to happen on 4% of my journeys.

Behaviour approaching traffic lights depends a lot on how well you know the lights. Generally people in cars accelerate towards green lights – because they can, and because it's annoying to miss a green light. But as a cyclist if you see a green light, as a rule of thumb it is better to slow down, especially if you see the lights in the distance, because the chance the light will still be green when you reach it is low. And it's safer, too.

50. ONE OF THESE DAYS

The quiz will be on Thursday. Here's one way to do the reasoning:

- If there were no true statements, E would be true, so we'd have at least one true statement, which is a contradiction.

- If there were one true statement, E would be true. However, at least one of A, B and D must be true, since they cover the whole week, so we'd have at least two true statements, which is a contradiction.

- If there were two true statements, E and D would be true, so B would have to be false, making C true, so we'd have at least three true statements, which is a contradiction.

- So there are at least three true statements (meaning no more than two falsehoods).

- Looking at the claims about days, at most two of A, B, D and E can be true, and we're only allowed two lies, so two must be true and two false. We can therefore eliminate Friday and Wednesday, which are each only mentioned once.

- If the test is on Tuesday, B, C and D are all false, and if it's on Monday, A, C and E are false. That leaves Thursday as the only possible day. B, C and E are the three true statements.

This puzzle, if it were in fact a real note from a teacher, could put the students in a slightly precarious position, due to the nature of self-referential logic statements like this. As an example, during a lecture by the late mathematician and author Raymond Smullyan, he was testing the audience on their skills with propositional logic. At one point, he produced two envelopes and explained that one of the envelopes contained a dollar bill and the other was empty. The envelopes each had writing on them, and he challenged an audience member to see if he could work out which envelope contained the dollar.

The first envelope said: 'Of the two statements, at least one is false.' The second envelope's message was: 'The bill is in this envelope.' After some thought, the volunteer decided that the first envelope's message couldn't be false, for then it would be true. So it must be true, and the second envelope's statement is false. With confidence, he chose the first envelope.

To the audience's confusion, this envelope was empty. The other envelope contained the dollar. After being playfully accused of trickery, Smullyan assured his audience that he never lied. In fact, the second envelope was a true statement, and the first envelope was neither a true nor false statement. It was the audience that made the assumption that the statements on the envelopes must be either true or false, but reality doesn't always conform to such notions.

So perhaps Mr. Gordon will surprise his entire class with a quiz on Monday, but we'll assume he's a congenial sort.

Chapter 6 **MIND GAMES**

51. CATCH UP 5

Player A can be sure of winning by starting with stack 3. At the end of the game, the combined heights of the towers will be 15, so if either player reaches a height of at least 8, they guarantee a win. If A plays 3, either B includes 5 in their first move (by playing 2, 5 or 1, 5, or 5), in which case A can win (by playing 1, 4 or 2, 4 or 1, 4 respectively), or B doesn't include 5, in which case A plays 5 next move and wins. If A plays 1, 2, 4 or 5 as a first move, B has a response that can guarantee a win.

This simple adding game has a remarkably short history. It was invented by a group of academics at New York University in 2015. It's a very easy game to play – a six year old can understand it. Yet the tactics are far from obvious. In a game where there are ten stacks of height 1 to 10, a six year old could play an expert and still have a chance of winning. And the catch-up nature of the game means that whoever is behind after one move is by definition level or ahead after the next, so both players still feel like they are in with a chance until the final stages.

Like Diffy (see Puzzle 40), the game of Catch Up has turned out to be a game of remarkable hidden depths. For some values of N, for example, $N = 5$ or $N = 6$, it is known that player A can force a win; for other values (e.g. $N = 9$ or $N = 10$) player B can force a win. And in some games such as $N = 7$ or (trivially) $N = 3$, either player can force a draw with optimal play. But it's still not known if the game favours A or B overall. The strategy for high values of N, e.g. $N = 50$, is chess-like in its complexity, and it's still not known which player is favoured.

What this means is that if you were to play against a device with Artificial Intelligence, you would have a chance of winning. Although

AI machines can learn to win at chess, it's much harder for them to learn a winning strategy for Catch Up. In chess, if the leader gains an early advantage, the gap tends to magnify. In Catch Up, on the other hand, the leader can't gain a clear advantage until the very end of the game since (by the definition of the game) the other player can catch up and overtake after every move bar the final one.

52. TAKING THE BISCUIT

Alpha should take one chocolate biscuit, leaving seven chocolate biscuits and four lemon. The key to the game is ultimately to leave your opponent with two biscuits in one jar and one in the other – whatever move they make, you can follow with a winning move. From seven and four, Betty's next move can't be three chocolate biscuits (as Alpha would win straight away). Whatever other move Betty makes, Alpha will be able to follow it either by getting to five and three (and thereafter to two and one, or zero), or to two and one. Whatever happens, she has a guaranteed path to win.

This biscuit-taking challenge is known as Wythoff's game (named after the Dutch mathematician who invented it early in the 20th century.) The game has some remarkable properties. One of them is that it can played visually by moving a queen on a 9 by 9 chess board.

Think of the number of biscuits left in the two jars as being the co-ordinates of a queen on the grid.

The number of biscuits in the left jar is the X (horizontal) co-ordinate and the number in the right jar is the Y co-ordinate. The bottom left square of the board is (0,0). For example, if there are eight biscuits in the left jar and four in the right, the current position of the game is at (8,4) as shown below.

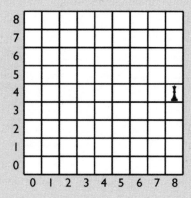

When it's your turn, you either remove from the left jar (which means moving the queen horizontally to the left) or from the right (queen moves down). If you remove equal numbers from both piles, that's equivalent to the queen moving diagonally. So suddenly we are playing a kind of chess, in which the aim is to get to the bottom left-hand square.

We've already seen that if you get to (2,1) or (1,2) you are in a winning position, so we can call those positions 'safe'. Before that come (5,3) and (3,5), then (7,4) and (4,7), and so on. Plotted out, the safe points lie along two distinct lines with a remarkable property: the slopes of the lines are the golden ratio (roughly 1.618) and its reciprocal (0.618).

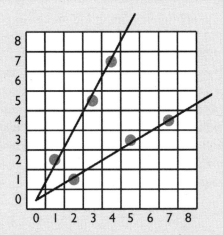

The golden ratio is a number with many links to the natural world (it is found in sunflower spirals, for example) and has also been much revered for its beauty throughout history, by Leonardo da Vinci among many others.

53. MEXICAN STANDOFF

There is slightly less than a 7/9 chance that Bumbling will survive.

The best strategy for each shooter is to take out the strongest rival first, so Good should aim first at Bad, while Bad and Bumbling should aim at Good. The chance that Bad will get through the first round is only 1% – so we'll be close to the right answer if we call that zero. Good only survives if Bad and Bumbling both miss him, and the chance that will happen is 1/3 × 2/3 = 2/9. If Good does survive, then he'll take out Bumbling in the second round. Therefore the chance that Bumbling will survive is 1 – 2/9 = 7/9, or 78%. The 1% chance of Good missing Bad reduces Bumbling's chances of survival very slightly.

This puzzle is of course a thinly disguised version of the iconic scene from *The Good, The Bad and The Ugly*. In that scene, the Good (played by Clint Eastwood), Bad (Lee van Cleef) and Ugly (Eli Wallach) form a triangle in a clearing in the middle of a graveyard, waiting for the first to pull his gun. If we assume that Clint Eastwood is the best shot because, well, that's the rule with heroes, then the maths suggests that Good and Bad should aim at each other, with Ugly also aiming at Good.

Spoiler alert – this isn't quite what happens. Good and Bad do indeed aim at each other, but Ugly chooses to aim at Bad. Good hits (of course), Bad misses (not surprising) and Ugly also fails to hit, though in his case it's because his gun doesn't fire. Good has removed the bullets from Ugly's gun.

The Good, Bad and Ugly shootout is better described as a three-way duel (a 'truel') rather than a Mexican standoff, as the latter is normally used to describe situations where at least two parties are already pointing guns at each other and nobody has a chance of escaping without serious injury. Genuine Mexican Standoffs are to be found in many movies, particularly those directed by Quentin Tarantino, such as *Reservoir Dogs* and *Pulp Fiction*.

54. MY FAIR LADYBUG

You can win in four guesses (at most) if you guess B, C, C, B (or C, B, B, C). If the bean isn't there when you first guess B, it will be under either B, C, or D next round. If your second guess of C turns up empty, then it will either be under A or C next round. If your third guess of C still doesn't turn out, the only place the bean can be on the fourth round is under cup B.

This puzzle is often presented as a story of a knight or a prince who knocks on one of four doors in order to marry a princess who inexplicably sleeps in an adjacent room each night, but we felt a thematic update was in order.

The puzzle is just as doable with five, six, or even one hundred cups, but some additional insight is necessary. The solution to five cups would be B, C, D, D, C, B. For six cups, it's B, C, D, E, E, D, C, B. Notice a pattern? As the bean moves, it will always swap from a cup in an odd-numbered position to an even-numbered position, or vice versa. When you start with the second cup and gradually move up to the second-to-last, you will certainly find the bean if it originally started in an even-numbered cup. Then, when you again choose the second-to-last and make your way back to the second, you will certainly find the bean if it originally started in an odd-numbered cup. This becomes clearer if you line up actual cups and try to intentionally 'dodge' the guesses.

55. DUNGEONS AND DIAGRAMS

Bernie starts in Room G, and ends in Room E, where he puts the treasure. There are four rooms with an odd number of doors: C, E, G and H. Since he goes through each door only once, he must start and end in two of those rooms, and the other two rooms must be connected by the secret door (making their total number of doors even). The only crate that joins two of those rooms is the one between C and H, so that must be the secret passage, and you start and end in G and E. Since we know that he places the chest next to a table, it must be in Room E, and he starts in G. There's more than one way he can achieve this. For example, starting in G his route could go:

G–F–C–A–B–D–C–G–H–C–D–E–B–H–E

This puzzle is a nicely disguised version of the famous Koenigsberg Bridges problem that was first posed and solved by the mathematician Leonard Euler in the 18th Century. The old city of Koenigsberg was built either side of a river that had a couple of islands in the middle. The different parts of the city were connected by seven bridges. On a Sunday stroll (so the story goes), Euler found that it wasn't possible to traverse all of the bridges exactly once, he needed to cross at least one bridge twice if he wanted to do the full tour. Euler had inadvertently opened up the new field of graph theory, the maths of networks.

One of the simple discoveries in this field was that if you are travelling around a network of 'nodes' connected by paths, it will only be possible to do this without repeating a path if the number of paths leading out of each node is even. The only exception is if there are exactly two nodes that have an odd number of paths leading out of them, in which case the journey needs to start at one of the odd nodes and finish at the other.

56. ANT ON A TETRAHEDRON

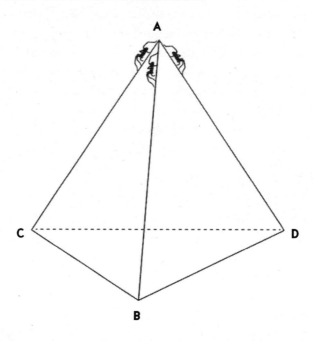

Two of the spiders each patrol up and down a wire, the first up and down AD, the second along BC. The four remaining wires, AC, CD, DB, BA form a connected path that the third spider can keep going round, eventually catching the slower ant. To prevent the ant from darting onto wire AD or BC (the ones being patrolled by the first two spiders), the first two spiders should co-ordinate their patrols so that they reach the end of their wire at the same moment as the third spider walks past it.

When this puzzle was first submitted, we convinced ourselves that there might be a way in which the ant could escape regardless of the spider strategy. It takes a clear head and an appreciation of 3-D geometry to find the solution.

Ants crawling over the edges of polyhedra have a long history in mathematics problems – they help to make problems more palatable. For example, the ability of ants to walk up walls and upside down on ceilings makes them ideal for picturing the shortest distance between two points in a rectangular room.

There's a vintage ant and string puzzle that is about adding an infinite series of fractions. In one version of the puzzle, an ant that walks at 1 cm per second is put onto a rubber string that, to begin with, is one metre long. Every second, the string is stretched instantaneously by exactly one metre. The puzzle asks whether the ant will ever reach the end of the string, and the surprising answer is that it will, but only after several light years. Poor ant!

Mathematician Sam Hartburn wrote a song called *The Ant's Lament* in sympathy with ants that are subjected to enforced involvement in mathematical puzzles. The chorus goes:

'Oh mathematician, why'd you do it, why'd you take me from my home
Leave me stranded on this rubber band, confused and all alone
Separated from my fellows and their guiding pheromones
An ant isn't born to be alone, an ant isn't born to be alone'

57. CHOPPING BOARD

The easiest way to solve this is to start from the end. If the board gets down to the two most senior members they will certainly survive, as member 2 (numbered from most senior to newest) will vote to stay on the board. Thus, member 3 will not be happy if the board comes down to just members 1, 2 and 3, because members 1 and 2 would then vote 'aye' and eject member 3.

It follows that with just members 1, 2, 3 and 4 left, the board would be stable; member 3 would vote 'nay' to prevent reduction to size 3, and member 4 would vote 'nay' to avoid immediate ejection. Similarly, members 1 through 8 would constitute a stable board, because if any of 5 through 8 voted 'aye' the board would quickly reduce to members 1 through 4. Since members 1 through 8 are a majority and are safe from further reduction, their 'ayes' will eject members 9 and 10, leaving 8 as the ultimate size of the board.

The concept of a vote among perfectly logical voters might be wishful thinking, but it does inspire some interesting and counterintuitive puzzles. Another popular puzzle with a similar setup is called the Pirate's Game. Five pirates have recently come across 100 gold coins. The most senior pirate can come up with a plan to distribute the coins among the crew, and his proposal will be voted on (the senior pirate votes also). If 50% or more of the votes are in favour, the coins are distributed accordingly. If not, the senior pirate is thrown overboard and the next most senior pirate must come up with his own plan, to be voted on the same way. The question is: what plan must the senior pirate propose such that he gets as much gold as possible while staying alive, assuming that the pirates are all perfectly logical in their deductions?

The answer: he can safely take 98 of the coins for himself and only propose to give one coin each to the third most senior pirate and the last pirate. The reasoning is similar to Chopping Board: if it were down to the last two pirates (D and E), D would propose to take all of the coins and leave none for E, as he only needs his own vote to enact the plan. So if there were three pirates left (C, D, and E), then C would propose 99 for himself and one for E, knowing that E would realize he would be better off with one than none, and thus vote in favour. Continue that logic backwards, and pirate A gets 98 coins with both C and E voting in favour for only one coin each. How's that for cut-throat?

58. ÉCLAIR-VOYANCE

Amy had the six, and Tom had the five. When Tom first says he doesn't know whether he'll win, this means he must not have either the two or the ten, or he would know for certain whether he had won or lost. Amy picks up on this, so when she says 'Same goes for me,' she implies not only that she doesn't have the two or ten, but also that she doesn't have the three or nine, for then she would know whether she had won or lost. Tom's admission that he still doesn't know means that he has neither the three or nine, nor the four or eight. So he must have the five, six, or seven. Amy's subsequent lack of knowledge means that she must have the six, since that is the only card where the outcome would still be uncertain. Realizing this, Tom admits defeat, meaning he has the five.

This a great example of a popular genre of 'no knowledge' logic puzzles, in which one person stating 'I don't have enough information' is in itself useful information. Perhaps the simplest, most elegant example of this is the black–white hat challenge. A teacher asks two students, Alice and Bob, to sit facing each other, and they are then blindfolded. The teacher announces that she has two black hats and one white hat, and (without revealing the colours) now places a black hat on each student's head, keeping the third hat hidden. The blindfolds are now removed, and the students are asked if they can deduce the colour of the hat on their own head.

Each student can see a black hat. This means that their own hat could be white or black, and they cannot be certain what colour it is.

When asked if they know for certain what colour hat they are wearing, both students say 'no'.

When this exercise is done on real people (adults or children) that's usually it. The participants just sit looking bemused. Yet both participants are in a position to make a confident deduction, simply by asking themselves: 'what if the hat on my head were white?' Let's suppose Alice's hat is white. Bob can see a white hat, and would know immediately that his own hat must be black. The fact that Bob says nothing means that Alice can confidently state that her hat is black. The same applies to Bob. As Sherlock Holmes might have put it, this is a case of the dog that didn't bark.

59. THE GOBLIN GAME

First of all, common sense says that Beth should place the Goblin on a lower number than Annie, because a smaller number will be reached before a larger one. But there is a particular choice of number that maximizes Beth's chances of winning, and that is to place her Goblin on square 6.

The average score on a dice numbered 1 to 6 is 3.5, so in the long run the chance of a counter landing on a square is going to be 1 in 3.5, which is about 29%. In fact, this turns out to be roughly the chance of landing on any square numbered 10 or higher, so for example there's an almost identical chance of eventually landing on the Goblin whether it's on, say, square 15 or square 31 or square 96.

However, for the low-numbered squares, the odds of landing on the Goblin are different. If the Goblin is placed on square 1, the only way to land on it is by rolling a 1 (a 1 in 6, or about 17%, chance). Square 2 can be reached in two ways, by rolling a 2, or by rolling 1 followed by another 1, which works out at just over 19% - slightly more likely than landing on square 1.

As you move along the squares, 3, 4, 5, ..., the number of ways of reaching that square increases, which means the chance of landing on it increases too. By square 6 the chance of landing works out to be about 36% when you add together the odds of each of the different combinations that could take you there. However, for squares 7 and above, things change. These squares can only be reached by landing on one of squares 1 to 6 first, and this means the chance of landing on 7 must be lower than the chance of landing on 6. Indeed, the odds of landing on a number higher than 6 turn out to always be below 30%. In other words, the best place to put the Goblin is square 6.

Since the outcome of a dice throw is random, and every number from 1 to 6 is equally likely to crop up, it does at first feel like there will be an equal chance of landing on every square. This puzzle shows that this is not the case, though after the early squares, the chance of landing on any particular square does rapidly converge on 1/3.5.

In games involving two dice, there is much more variation in the chance of landing on different squares. Although a roll of two dice can score totals of between 2 and 12, the chance of rolling a score of 7 is six times as high as rolling either 2 or 12.

The most familiar two dice game is probably Monopoly, and mathematicians have investigated which are the most visited squares on the board. Most visited is the 'Jail' square, which is not really surprising since it is represented by two squares: 'Go To Jail' and the Jail itself. This bias towards landing on Jail affects other squares. With rolls of 6, 7 or 8 being the most likely with two dice, the two orange squares either side of Community Chest get more than the average number of visits, which makes them the best investment on a Monopoly Board.

60. WEATHER OR NOT

Eileen has a better chance of winning. Call a Rain day R and a No Rain day N. Eileen chooses NRR and Joe RRN. In the first three days, they each have an equal (1/8) chance of being right. But if the sequence RRN first crops up after day 3, then NRR must have occurred a day earlier unless it rained on every previous day. So for Joe to win, the sequence has to be RRN or RRRN or RRRRN or ..., for which the probability is 1/8 + 1/16 + 1/32 + ..., which adds to 1/4. The chance of Eileen winning is therefore 3/4, three times as likely as Ike.

This puzzle is a nicely disguised version of the classic game of 'Penney Ante'. In Penney Ante (devised by the mathematician Walter Penney in the 1960s) you take a coin, and challenge somebody to come up with a sequence of three flips of a coin (for example Heads–Heads–Tails, or HHT). You then choose your own sequence (in this case you should choose THH for reasons that will be explained shortly), and bet that if a coin is now fairly flipped until one of the two sequences comes up, your sequence will come up before theirs.

It is natural to suppose that since this involves flipping a coin, and every outcome is equally likely, you each have an equal chance of winning the bet. But this turns out not to be the case. The easiest example to illustrate why is when your opponent chooses HHH. Let's suppose you keep flipping a coin until HHH appears. The very fact that it is appearing for the first time means that the previous flip must have been a Tail. If it had been a Head then that would have been the first time the sequence HHH appeared. So THH will always occur before HHH, except when the first three flips are HHH. The chance of throwing three Heads in a row is 1/2 × 1/2 × 1/2 = 1/8, which means HHH will win 1/8 of the time and THH will win the other 7/8 of the time.

For every sequence of three Heads and Tails, there is always another that will beat it the majority of time. The rule for finding the winning pattern is to take your opponent's choice (HTH, say), delete the last coin (leaving HT) and then add at the beginning whichever H or T makes your pattern asymmetrical. In this case you'd go for HHT, because THT has symmetry.

61. THE CAKE AND THE CANDLES

The candles and the knife cut are all at random points along the length of the cake. Two of the lines in the diagram below pass through the points where the candles are placed (C) and the other is the line along which Victoria cuts (V). We can't tell which is which.

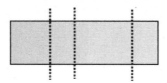

The order of the lines is therefore one of:

C C V

C V C

or   V C C

The cut only divides the candles in one of these three possibilities, so the chance of there being a candle on each piece is 1/3.

In its original form, the candle problem was posed in terms of making two randomly placed notches on a stick and then breaking the stick at some random point. What (the puzzle asked) is the chance that there is a notch on both pieces of stick? Although this is arguably the same puzzle, there's an important practical difference. How do you break a stick at a random point? It is very hard to break a stick at a point close to the end, so if you happened to choose a point (say) 5 mm from one end, you'd probably need a hacksaw to create two pieces. The storyline was therefore changed to being a cake, which is more easily cut, and with candles instead of random marks to give a whimsical purpose to the whole exercise.

The most common wrong answer to the puzzle is that the chance is 50–50. One (plausible but fallacious) line of reasoning is that the cake has a left half and a right half, and whichever half the first candle is on, the chance that the second candle will also be on that half is 50%. On average the cut will be in the centre so, the argument goes, there's a 50% chance that both candles will be on one side of the knife, and a 50% chance they'll be either side. The problem with probability puzzles is that there are often answers that sound plausible but which turn out to be wrong. This is an example of one of those puzzles.

There is another classic stick breaking puzzle that was made famous by the great American maths popularizer Martin Gardner. The puzzle asks the apparently simple question: if you break a stick at random into three pieces, what is the probability that the three lengths can form a triangle? This puzzle might sound similar, but involves some very different thinking. It also depends on how the random breaks to the stick are made. If you break the stick at some random point and then choose one of the two pieces at random and break that at a random point, the chance of forming a triangle turns out to be 1 in 4 (or 25%). But if you pick two random points first (as in the original puzzle above) and then break the stick at those points, the chance of a triangle is 1 in 6 (or 17%).

62. VIVE LA DIFFÉRENCE

Swapping seats 3 and 5 (to give the order 1 2 5 4 3 6 7) increases the discount to 24 euro. This is the maximum possible discount, and although there are several other ways of arranging the chairs to get a discount of 24, this is the only one that involves swapping only two chairs.

The solution can be found with some intelligent experimentation. For example, before the seats are moved, the neighbours of seat 6 are seats 7 and 5, which only differ by 2. That seems like a 'waste' of the 7; it would help if it were two away from a smaller seat number. Likewise, seat 1 is two away from seat 3, with a difference of only 2, so it would help to replace chair 3 with a bigger number. Swapping chairs 3 and 5 turns out to be the sweet spot that maximizes the differences.

When Radio 4's *Today* programme announced it was visiting Newcastle University, mathematicians were invited to come up with ideas for the programme's daily puzzle. Christian Lawson-Perfect put forward an early version of this restaurant puzzle as his entry, but alas it was rejected. In that first version, there were only five customers in the restaurant and they were trying to minimize their total bill. The puzzle was a little unsatisfactory in that the best solution with five guests turns out to be seating the customers in the order 1 2 3 4 5, giving a difference total of 12.

We encouraged Christian to revive the idea for *New Scientist*, and see if there was a version of the puzzle with a more interesting solution and storyline. And there was! It turned out that the problem becomes richer when the number of customers increases to seven. 1 2 3 4 5 6 7 is not the optimal arrangement, and swapping chairs 3 and 5 produces the highest difference value. We needed a credible storyline, and came up with an eccentric restaurant owner. Now all that was needed was a name for the restaurant. Ideas included Bistro Subtraction, Café Differo and Chez Absolue, but by far the best was Christian's suggestion: Vive la Différence.

63. DIAMONDS ARE FOREVER

There were no diamonds in the box. Call the number of palaces P. There were P^2 vases, and P^3 diamonds, of which Fidelio got P, leaving $(P^3 - P)$ for the daughters. Whatever value you choose for P, you'll find that $P^3 - P$ is exactly divisible by 6. Why? Because it is equivalent to $P \times (P + 1) \times (P - 1)$, the product of three consecutive numbers, one of which must be divisible by 3, and at least one divisible by 2, making the product divisible by 6.

This puzzle has appeared in at least two previous guises in *New Scientist* over the years, each time with an ancient tale about stored treasure, and exploiting the fact that the number would always be divisible by six. This version was no exception, but it added the extra twist of asking the reader to work out the exact number of diamonds in the box without being given any information about how many diamonds there had been originally.

64. CHANGING THE GUARD

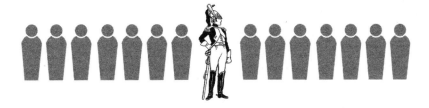

Perkins will end up facing right, regardless of where the other left-turners were standing. The five soldiers who are facing left to start with will 'transfer' that leftness to a soldier to their left until there are no more right-facers to transfer to. The soldiers will end up arranged like this:

L L L L L R R R R R R R R

Perkins is eighth from the left, so will end up facing right.

This is an example of a parity puzzle (these are puzzles based around sequences of odds and even numbers). It has the same underlying logic as another puzzle that appeared in *New Scientist*. That puzzle was about six so-called peg beetles on a clothes line, pointing left or right, including two that were at the end of the clothes line pointing inwards.

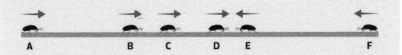

The peg beetles march at a steady speed, reversing direction whenever they meet another beetle, and fall to the ground when they get to the end of the clothes line. The challenge is to work out how long it will be before all the beetles have fallen off the line. Unlike the regimented soldiers in the parade puzzle, the interactions of the beetles appear more chaotic and random, making it harder to work through a solution. But if you blur your eyes, two peg beetles meeting each other and reversing direction looks just the same as the two beetles walking past each other. Therefore the time it takes for all the beetles to fall off the line is simply the time it would take for a beetle at one end of the clothes line to walk to the other end.

65. YAM TOMORROW

Abel, Babel and Cabel received 10, 7 and 5 yams respectively, and the monkey 3. Suppose in the morning they find 3 yams and each get one. This means in the night Cabel took $(1.5 \times 3) + 1 = 5.5$ yams – but we need it to be a whole number. Try increasing the number they get to 2 each (6 yams): this means Cabel found 10 yams (and took 3), Babel found 16 yams (and took 5) and Abel found 25 yams (and took 8). The next solution with whole numbers would be 106 yams, but that exceeds the capacity of the crate.

This is a more accessible version of a puzzle that first appeared in a magazine called *The Saturday Evening Post* in October 1926. The puzzle was in a story with the title 'Coconuts' by Ben Ames Williams. The story involved a contractor distracting his competitor with a puzzle so exasperating, it caused the competitor to miss the deadline for entering his bid. The puzzle is very similar to Yam Tomorrow: there are five sailors shipwrecked, and during the night, each sailor attempts to divide a pile of coconuts into five equal piles, but there is one coconut left over, which is given to the monkey. They then take one of the piles to hide it for themselves. After each sailor does their division, they awake in the morning and finally divide the remaining coconuts equally among themselves, with nothing left for the monkey. The question was, 'How many coconuts were there in the beginning?'

The setup wasn't new. A similar puzzle featuring four brothers, a monkey, and a pile of nuts on a table even appeared written in the journal of author Lewis Carroll in 1888. In fact, the basic mathematics are traced back to Diophantus of Alexandria, an ancient Greek algebraist, and specific problems in this vein were discovered as early as 850 CE in the work of Indian mathematician Mahāvīra, who dealt with the continual division of fruit and flowers with specified remainders. But the specifics of Williams' puzzle are notable in their difficulty and the way they captured the public imagination. Unlike the puzzles that appear in *New Scientist*, the solution was completely unknown to *The Saturday Evening Post*. Over 2,000 letters were sent in within the first week of publication, causing the editor-in-chief to send Williams a wire that read: 'FOR THE LOVE OF MIKE, HOW MANY COCONUTS? HELL POPPING AROUND HERE.' Even twenty years after the publication, Williams was still personally receiving letters requesting the answer or proposing solutions.

The coconuts problem can be written out and 'simplified' to the following equation: $1024N = 15625F + 8404$, where N is the number of coconuts at the beginning and F is the amount of coconuts each sailor receives in the morning. This is called a Diophantine equation (after the Greek mathematician), which are equations that often have two or more unknowns and are looking for solutions in whole numbers. However, an equation of that size is much too complicated to be solved through trial and error, and the analysis to get the answer is fairly involved. In fact, there are an infinite number of solutions in whole numbers, but the task is to find the smallest such answer. Over the years, there have been a number of ingenious methods that shortcut the usual lengthy process, from utilizing 'negative coconuts' to adding special blue coconuts to using a base-5 counting system, all of which have been used to arrive at the correct answer.

What is that answer, you ask? Perhaps you should write a letter to Mr. Williams and ask him. (Ok, it's actually 3,121.)

66. BLURRI-NESS

The correct answer is 15 metres.

When the boat reaches the ripple (point *X*), the ripple has been travelling for 5 seconds; when the boat crosses the other side (*Y*), the ripple has been travelling for 15 seconds, so the diameter of the outer ripple in the diagram is three times the inner ripple, meaning the ripple originated 1/4 of the way between *X* and *Y*. The boat took 10 seconds to cross the ripple, so they reached the monster 2.5 seconds after crossing the ripple or 7.5 seconds after taking the photo. 2 m/s × 7.5 = 15 m.

The original storyline for this puzzle featured a couple of characters who spot a fish while boating on a river. We needed more jeopardy. Fortunately it was a short step to turn it into a Loch Ness story. The puzzle was published in the spring of 2022. A week later, in what felt like more than just a coincidence, there was a news report about Tom Ingram, a tourist on a Loch Ness boat trip, who had spotted 'something big' appearing on the boat's sonar. In Ingram's case it was the boat's sonar and computers that did the calculations. It turned out that the mysterious moving object was located at a depth of around 400 feet and was about 30 feet long.

The sonar image was screen-grabbed by Mr Ingram and published widely in the media. Needless to say, Ingram's image was somewhat blurry. (Aren't they all?)

The *New Scientist* puzzle appeared on 26th March and the sonar sighting was on 4th April, with a certain Fool's Day almost midway between the two. The French might regard this as a case of Poisson d'avril.

67. NO TIME TO TRY

There are 24 ways of arranging four digits ABCD (such as ABDC, ACBD, etc.), but there are 36 ways of arranging three digits into a four-digit number (AABC, AACB, ABAC, etc.). Why 36? The repeated digit can go in six pairs of positions (AAxx, AxAx, AxxA, xAAx, xAxA and xxAA), and there are three digits that could be 'A', making 6 x 3 = 18 possibilities. The other two digits can be arranged either BC or CB, making 18 x 2 = 36 altogether. Blond should go for Door 1 with the four worn digits and he'll only need 24 attempts at most.

The puzzle originated when its setter Katie Steckles was on a walk and came across a worn keypad with only three numbers worn out. She wondered if the lock had a three-digit code, or if – as is more common – there was a four-digit code, which would mean one digit must be repeated. Her immediate (as it turns out incorrect) hunch was that there would be fewer ways to make a code using three digits than four. Katie submitted this puzzle just before the delayed launch of the James Bond film *No Time To Die*, and the storyline was given a 007 theme to make it topical. The punny title was particularly satisfying.

There is an interesting extension. What if you were to find only two worn digits on the keypad? This time your hunch is right if you reckon the number of arrangements will be reduced. There are only fourteen possible arrangements; eight in which one digit is used three times (ABBB, BABB, BBAB, BBBA, BAAA, ABAA, AABA, AAAB) and a further six with two of each (AABB, ABAB, ABBA, BAAB, BABA, BBAA). And of course if there is only one worn digit, there is only one possible combination: AAAA!

Next time you encounter a keypad with an unknown four-digit code, keep your fingers crossed that it doesn't have three worn digits.

68. PAINTINGS BY NUMBERS

The neighbouring rooms of Room 5 (say) must include Rooms 4 and 6, and in general the neighbours of any odd-numbered room must be even numbers, while even numbers must have odd neighbours. We can think of the gallery as a mini-chess board, with odds on white squares and evens on black squares. Since there are five odds and four evens, the corners and the centre square must be the five odd numbers. Where can Room 1 be? It can't be bottom left as the two rows above it must add to more than 200. By trying to trace paths 1-2-3... and so on, you can also quickly rule 1 out from bottom right, top right and centre. The only path that works is:

This puzzle was an adaptation of a *Tantalizer* puzzle from the 1970s about sheep in a field of square pens. It's not known if the originator of that sheep puzzle, Martin Hollis, was the first to discover this pattern of numbers with its unique addition property.

We updated the story to a setting in which rooms are typically numbered, and where it's normal to visit each room exactly once, to make the storyline *almost* feel plausible.

69. A PIAZZA OF DOMINOES

A full set of dominoes has eight of each number. In the piazza there are eight 0s and 1s, and seven 2s, 3s, 5s and 6s, so we can be sure that the 0-0, 1-1, 2-2, 3-3, 5-5 and 6-6 dominoes all appear somewhere. 6-6 and 0-0 can be identified immediately, which forces 5-6 (bottom left) and 5-3 (bottom right), so barriers can now be drawn between any other 5-3 and 5-6 pairs, and 5-5 must be in the top row. From this the whole pattern can be found. The missing dominoes are 4-4, 4-2, 5-4 and 6-3.

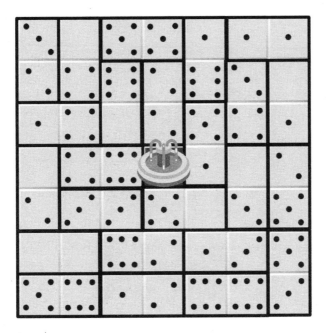

A 7 x 7 grid has 49 squares, so it's necessary for the fountain to use up a square in order to leave an even number of squares (48) for the dominoes to be placed on. Of course, having the fountain in the centre is aesthetically pleasing for its symmetry, but couldn't the fountain have just as easily been anywhere else, such as one square to the right or left?

To answer that question, we can look at a famous puzzle first posed by philosopher Max Black in 1964: if you were to remove two opposite corners on a chessboard, could you completely tile the remaining squares with dominoes, covering every square? If so, how?

The key to this puzzle comes from realizing that a domino must necessarily cover exactly one black and one white square – never two of one colour. So in order to successfully tile an area with dominoes, there would need to be an even number of black and white squares. In this case, two black squares were removed, so it will be impossible to cover the chessboard.

Back to our piazza, we can gain insight by shading the tiles in a checkerboard pattern.

In this pattern, there are 25 black squares and only 24 white squares. In order to tile completely with dominoes, we need an even number of each colour, so the fountain can only be placed on one of the black squares to leave 24 black and 24 white squares on which to place the dominoes.

70 TROUBLE BREWING

If you press the coffee and milk buttons you get... coffee and milk!

There's a quick way to the solution if you have an 'aha' insight.

First, think about what happens if you press all five of the buttons. You will get nothing. Each ingredient is being requested twice, and hence cancelling itself out.

What this means – and here is the 'aha' moment – is that the result of pressing a particular set of buttons will be the same as the result of pressing all of the other buttons instead. For example, when you press milk and sugar, you will get the same result as if you press the other three buttons: tea, chocolate and coffee.

We are told that pressing tea, milk and sugar delivers tea, milk, sugar and coffee. This means that pressing chocolate and coffee also produces those four ingredients.

But we are told in the puzzle that pressing chocolate and milk produces tea and sugar. Ordering coffee and milk is the same as ordering (coffee and chocolate) and (chocolate and milk). The first produces tea sugar milk coffee, the second cancels the tea and sugar, and leaves milk and coffee.

We find it quite amusing that, after going through all that effort to work out the logic behind the rewired machine, it turns out that coffee and milk still deliver coffee and milk. And it's here, as we reach the end of this collection, that we can reflect on what makes for a satisfying puzzle. Sometimes it's because of a pithy or humorous storyline. At other times it's an 'aha' moment that gives you insight into the problem or a shortcut to the solution. Or it could be that there's a surprising or entertaining solution that serves as a twist at the end or, dare we say, a 'haha' moment. Perhaps in this case, as with many of our favourite puzzles that we've come across, you push the button and get a little bit of everything.

PART 3

HINTS

Chapter 1 **PUB PUZZLES**

1. CREATIVE ADDITION

Since the columns differ by 3, the first idea is often to move – or remove – the 3 card, only to realize that it doesn't help. Think about what 'move' can mean, and think about what material the numbers are written on.

2. THE H COINS PROBLEM

One solution is, to use a hackneyed phrase, 'inside the box' and the other is 'outside the box'. But (unlike some of the solutions to Puzzle 1) there is nothing that might be regarded as cheating required here. In both solutions, there are never more than three coins making a straight line – but the spacing between coins varies.

3. THE BOOK OF NUMBERS

According to dictionary convention, 'Six thousand' appears before 'Sixteen' because a space (e.g. the one between the 'x' and the 't' in six thousand) comes before 'a' in the alphabet.

4. LATE FOR THE GATE

Imagine yourself and a friend approaching two parallel travelators. Follow through what happens when one of you stops to do your laces while the other walks to the travelator before doing so.

5. BUS CHANGE

You could have £1.01 in the form of a 50, two 20s, a 5 and three 2s and still not be able to make exactly £1. But that's not the maximum.

6. DARTS CHALLENGE

The highest scores are reached with a combination of trebles and a bullseye. 180 = 3 × 60, so the next achievable score below that is 177, and below that is 174.

7. EVENING OUT

How many ways can you find of making 9 even by moving or removing just one or two of its own matchsticks?

8. CAESAR CIPHER

What if you write 3 as the word THREE, and shift the letters along the alphabet?

9. SYMMETRIC-L

Nothing beats cutting out three Ls and playing with them. One solution looks a bit 'fishy', and you might *love* the other one (which requires you to flip one of the Ls over).

10. WHICH DOOR?

What was the chance that the prize was behind one of the three doors that Kelly first picked? And have those odds changed when the host opens doors that he knows have nothing behind them?

11. BONE IDLE

What would be the worst possible outcome if you were to study seven of the eighteen topics?

12. SUNDAY DRIVERS

Think about which driver is most likely to create a clump behind them, and which positions they could or could not be in.

13. LEAGUE OF NATIONS

Scotland didn't play in the first weekend so they must have played every other weekend. Could they have played their first game against Wales?

14. FASTEST FINGERS

Could they all have got one of the four answers in the right position? No, because that would mean exactly one of Jasmine's letters CDBA is correct, but whichever letter you decide is correct will lead to Virat or Finnbarr having zero or two correct.

15. RESHUFFLING THE CABINET

Since Dyer is moving on to be Health Secretary, neither Anerdine nor Brinkman can be moving to take Dyer's job.

16. AMVERIRIC'S BOAT

No difficult geometry is needed here. Imagine the boat moves in by a distance x and see what you can deduce from the triangle formed by the rope in its original and end position.

17. EXPRESS COFFEE

Suppose the depot is at A. What happens to the combined distance from the stalls to the depot if you move the depot from A to B?

18. SEVENTH TIME LUCKY?

If each column only contains one correct digit, then only four of her guesses will contain a correct digit. So you should look for columns where there are repeated digits.

19. THE TWO EWES DAY PARADOX

Suppose a friend flips two coins. There are four possible outcomes: HH, HT, TH and TT. In how many of these outcomes could your friend tell you 'at least one of these coins is showing a Head'?

20. THE HEN PARTY DORM

As soon as somebody occupies Ava's bed, everyone else will end up in their own bed. So if you arrive and find that your bed is occupied, it means Ava's and Janice's beds are still empty.

21. CUTTING THE BATTENBERG

The four solutions we're thinking of each result in a piece of cake that is a different shape. How can you slice a piece of Battenberg so that it has the same amount of yellow and pink sponge?

22. A JIGSAW PUZZLE

In a rectangular jigsaw that is 20 across by 15 down there are 300 pieces. How many rows and columns are needed for there to be 468 pieces?

23. THE NINE MINUTE EGG

If you want to minimize the time, any solution starts with flipping both timers. After this, it's about when to flip them again, and when the egg should start boiling. A slow solution will take 16 minutes from first flip to a boiled egg, but there's a way to do better than this.

24. MURPHY'S LAW OF SOCKS

After one wash there is bound to be exactly one odd sock. With two pairs plus a singleton, which sock needs to disappear if there are still to be two complete pairs?

25. REARRANGING BOOKS

There is exactly one arrangement of those books on the shelf that would require as high as nine moves to get back to numerical order. What is that arrangement, and what might that tell you about what to look for?

26. HIDDEN FACES

What do the opposite sides of a dice always add up to?

27. BIRTHDAY CANDLES

Try this with only four candles. How many blows would it take then, and which candles have to be aimed at?

28. LIGHTBULB MOMENT

At the top of the stairs you can find the first circuit. How can you then set yourself up to gain more information once you get back to the basement?

29. BLOXO CUBES

Playing with some toy bricks might help.

30. KNIGHT NUMBERS

A knight and a bishop won't ever type the same number, because knight digits alternate between odd and even, but bishop numbers never do.

31. SUM THING WRONG

If the first digit were greater than 4, the sum would add to a five digit number.

32. CAR CRASH MATHS

The energy of a moving object of mass m moving at speed V is given by the formula: $E = \frac{1}{2}mV^2$. The value of V^2 for the 100 mph car is 10,000 units and for the 70 mph car is 4,900 units.

33. SOCCERDOKU

Albion lost two matches, so they had a deficit of at least one goal in each game. And yet they have scored as many goals as they have conceded, which tells you something about their third match.

34. ALL SQUARES

The year 1936 was a square number. Which would be the next square year after that? Also, a reminder of some school algebra: $(a - b) \times (a + b) = a^2 - b^2$.

35. SQUAREBOT

If you ask Squarebot to square the width of the rectangle, and it says '289', that doesn't tell you if it's a 17 by 17 square or a 17 by 18 rectangle.

36. CHRISTMAS GIFTS

I receive five gold rings on eight days (from Day 5 through to Day 12).
That's 5 × 8 = 40 in total.

37. TIGHTWAD'S SAFE

Which row or column must 0 and 1 be in?

38. THE CARD CONUNDRUM

There's more than one direction that you can look at the card.

39. MARTIAN FOOD

There is some number of people that will eat exactly the amount that
will be replenished the next day. You only need to consider the extra
people above this number, as these are the ones depleting the stock.

40. DIFFY

If the starting numbers are evenly spaced, you get to a zero square
quickly.

41. PIECES OF EIGHT

In a digital zero, six of the seven bars are lit up. In a four, only four are lit up.

42. WHICH FLIPPING YEAR?

Wouldn't it be nice if the answer to this was a date with some historical significance?

43. SIX WEEKS OF SECONDS

The number of seconds in six weeks is $6 \times 7 \times 24 \times 60 \times 60$. Now break those numbers down into factors (e.g. $6 = 3 \times 2 \times 1$).

44. LARA'S BIRTHDAY

If you are born on 29th February, on what date do you celebrate your first birthday?

45. THE MOUNTAIN PASS

Think about distances as well as times. Bonnie will travel in 4 hours exactly what Aaron travelled in the morning, and Aaron will travel in 9 hours exactly what Bonnie travelled in the morning. Ratios are important.

46. A WELL-TIMED NAP

For every 60 minutes on the clock, the hour hand moves through one twelfth of a circle, which is 30°, so if the time is *m* minutes past the hour, then the hour hand has travelled through 1/2 *m* degrees past the hour.

47. TRIPLET JUMP

If Connie is 29 when she has the babies, how old is she when they are 15?

48. HALF TIME

Add up all the numbers from one to twelve and find half of that number.

49. SEEING RED

If the light turns green immediately when he first sees it, will he be able to make it? What's the furthest away he could be when the light turns green and still be able to make it through?

50. ONE OF THESE DAYS

Start by looking for a contradiction. For example, if statement A is true, what does that say about the truth of the other statements?

51. CATCH UP 5

Leading with a 5 will lose, as your opponent can play 1, 3, 4 to make 8 (which is a guaranteed win).

52. TAKING THE BISCUIT

What happens if you are left with two biscuits in one jar and one in the other?

53. A MEXICAN STANDOFF

Who is Good at greatest risk from, and where should he therefore aim?

54. MY FAIR LADYBUG

If you guess cup B on your first go and it isn't there, it must be at A, C or D, and hence at B, C or D next round. What if you guess C on your second go?

55. DUNGEONS AND DIAGRAMS

If a room has an odd number of doors, is it possible to start and finish in that room?

56. ANT ON A TETRAHEDRON

If two spiders each pick a different edge to 'patrol', the third can go on a continuous loop around the other four edges.

57. CHOPPING BOARD

The newest member will always vote to preserve themselves, so what would happen if there were four members left on the board?

58. ÉCLAIR-VOYANCE

Tom can't have 2 or 10 because he'd know he had definitely lost or won. This means Amy can't have 3 or 9 or she'd know from Tom's reaction that he had won or lost.

59. THE GOBLIN GAME

If the Goblin is placed on square 1, then the only way a player can land on it is by rolling a 1 (a 1/6 chance). But if the Goblin is placed on (say) square 3, it can be landed on by rolling a 3, or 1 then 2, or 2 then 1, or 1, 1 and 1, so there is a higher chance that a player will land on it.

60. WEATHER OR NOT

If, after a few days, the sequence goes Rain-Rain-Dry for the first time (and Ike wins), then the day before this it must also have rained or Eileen would have won with Dry-Rain-Rain.

61. THE CAKE AND THE CANDLES

From left to right it could be candle-candle-cut, candle-cut-candle or cut-candle-candle. Are these all equally likely?

62. VIVE LA DIFFÉRENCE

In most cases, swapping neighbouring seats, for example 3 and 4, doesn't change the discount. The exception is swapping seats 1 and 7, which actually reduces the discount to 16 euro. To increase the discount, the two swapped chairs must therefore be at least two apart from each other.

63. DIAMONDS ARE FOREVER

If there were P palaces, then there were P^2 vases, and P^3 diamonds, so the daughters have been left with $P^3 - P$ diamonds to share. Try this out with a few values of P. What do you notice?

64. CHANGING THE GUARD

If a soldier is facing left and there is somebody in the row to the left of them who is facing right, will they get to turn right at some point?

65. YAM TOMORROW

It might be easiest to work backwards. If each sailor gets N yams in the morning share, there were $3N$ left in the morning. So before Cabel took his third, there were $9N + 1$ (the monkey got the 1).

66. BLURRI-NESS

The ripple is, of course, getting wider at a constant rate. When the boat meets the ripple for the second time, how much wider is the ripple than when the boat first crossed it?

67. NO TIME TO TRY

If only three buttons are used (call them A, B, C) then the possible codes include BCAB, CAAB and many other combinations. Can you count them all?

68. PAINTINGS BY NUMBERS

Room 1 cannot be in the bottom left, but where could it be?

69. A PIAZZA OF DOMINOES

In a complete set of dominoes there are eight of each number (for example 1 appears on 1-0, 1-1, 1-2, 1-3, 1-4, 1-5 and 1-6). There are eight 0s and seven 6s in the piazza so the 0-0 and 6-6 dominoes must be there, and can be found immediately.

70. TROUBLE BREWING

When chocolate and milk are both pressed, only two ingredients come out, so they must have an ingredient in common which cancels out. What must it be?

CONTRIBUTOR BIOGRAPHIES

Catriona Agg is a maths teacher who is well known for the elegant geometrical puzzles that she poses on Twitter and elsewhere.

David Bedford is an Honorary Senior Lecturer in Mathematics at Keele University.

Donald Bell is a former Director of the National Engineering Laboratory and is now active in a variety of maths groups.

Holly Biming: Please see the Acknowledgments.

David Bodycombe is a puzzle author and games consultant who has worked on several TV game and quiz shows including *The Crystal Maze* and *Only Connect*.

Derek Couzens is a retired maths teacher who works part-time for the Advanced Maths Support Programme, and full-time as a taxi driver for his grandchildren.

Paulo Ferro is a maths teacher and author who posts regularly on his website ENIGMATHS: https://en1gm4th5.wordpress.com.

Chris Healey has written two puzzles for *New Scientist* separated by a gap of 40 years, during which time he's squeezed in a career as a sound engineer.

Hugh Hunt is Professor of Engineering Dynamics and Vibration at the University of Cambridge, Keeper of the Clock at Trinity College Cambridge and Deputy Director of the Centre for Climate Repair.

Andrew Jeffrey is a puzzle-loving teacher, podcaster and presenter, as well as the founder of Maths Week England. www.andrewjeffrey.co.uk.

Christian Lawson-Perfect is a learning software developer at Newcastle University and recreational mathematician. https://somethingorotherwhatever.com.

Chris Maslanka is a writer and broadcaster who specialises in puzzles, his Pyrgic Puzzle column has appeared weekly in the *Guardian* for over 20 years.

Alex Mayall and **Alaric Stephen** are a venture capitalist from London and a sixth form maths teacher from Worcester respectively. Together they host the mathematical podcast Odds and Evenings.

Zoe Mensch represents many people: please see the Acknowledgments.

Peter Rowlett teaches maths, including the maths of puzzles and games, at Sheffield Hallam University and is a podcaster and blogger on mathematical topics. https://peterrowlett.net/.

Ben Sparks is a mathematician, teacher and maths communicator who talks and writes about maths around the UK (and the rest of the world). www.bensparks.co.uk.

Katie Steckles is a mathematician, lecturer, writer and presenter based in Manchester. www.katiesteckles.co.uk.

Chris Tiernan was an experienced investment consultant and a keen recreational mathematician.

Angus Walker is a maths graduate, infrastructure planning lawyer, puzzle setter and was the winner of the GCHQ Puzzle Book competition.

Howard Williams is a retired accountant still maintaining his balance in West Wales.

Peter Winkler is a Professor of Mathematics and Computer Science at Dartmouth University, and has written three books of mathematical puzzles.

ACKNOWLEDGMENTS

Writing puzzles is often a team effort, and it's rare for any *New Scientist* puzzle to arrive fully formed. It usually takes a second pair of eyes to spot wording ambiguities, confusing solutions, or clunky stories that are just a bit 'meh'. Many of the puzzles in this book were the work of more than one person. Zoe Mensch has been a stalwart setter, representing (as she does) 'zwei Menschen', two people, though rarely the same two people. The Zoe Mensch pairs have included at different times Howard Williams, Katie Steckles, David Bedford, Alison Kiddle, Zoe Griffiths, Catriona Agg and Matt Scroggs - plus ourselves and a few others, too.

We'd like to give huge thanks to the many checkers who role play as *New Scientist* readers, and who give a thumbs up or thumbs down to draft puzzles. Special thanks to Kyle Evans and his sixth form Further Maths classes who have made *New Scientist* puzzle checking a feature of their Friday afternoon lessons down in Winchester. Other invaluable checkers have included Colin Wright, Cath Moore, Mary Ellis, Nicole Cozens, Ben Sparks, Tom Rainbow, Christian Lawson-Perfect, and not forgetting, closer to home, Jenna, Geoff and Tom Eastaway.

In compiling this collection, we are indebted to Stephen Power, who not only read through various drafts, but also provided valuable insights on which puzzles we should include.

For the copy-editing we were also fortunate to be able to tap into the expertise of Sam Hartburn and Colin Beveridge who between them made numerous corrections and improvements that we would never have thought of.

Some of the puzzles in this book have drawn their inspiration from Tantalizers and Enigmas that appeared in *New Scientist* in the 1970s and 1980s. The outstanding puzzle setter in that era was Martin Hollis, who produced witty, whimsical logic puzzles week after week for many years. The current *New Scientist* column has sometimes tapped into Martin Hollis' archive, adapting the character names, titles and stories of some of his long forgotten puzzles to fit in a modern setting. Holly Biming is a reincarnation of Martin Hollis, but his influence is to be found in other puzzles as well, including some of our own. The legacy of *New Scientist* puzzle-setting stalwarts Eric Emmet and Stephen Ainley is also to be found in these pages.

We'd like to thank the team at Atlantic for all their hard work behind the scenes - Ed Faulkner for setting everything up, Kate Ballard our editor, and Rich Carr the typesetter.

We're also grateful to Cat de Lange who helped to facilitate the three-way dealings between publisher, magazine and authors.

Finally, our thanks also go to Julia Brown, who got the column up and running and was its guiding hand for the first twelve months, to Tim Revell, who enthusiastically picked up the baton and ran with it, and to Alison Flood, who brought in a fresh new perspective to the column. All three have been a pleasure to work with.

Rob Eastaway & Brian Hobbs

NOTES